Changing Core Mathematics

Library of Congress Catalog Card Number 2002101381

ISBN 0-88385-172-5

Printed in the United States of America

Current Printing (last digit):
10 9 8 7 6 5 4 3 2 1

Changing Core Mathematics

Edited by

Chris Arney
The College of Saint Rose
Albany, NY

Donald Small
United States Military Academy
West Point, NY

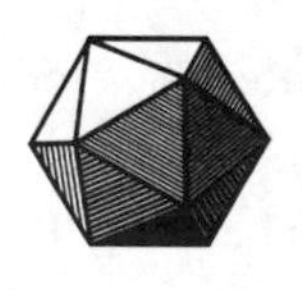

Published and Distributed by
The Mathematical Association of America

The MAA Notes Series, started in 1982, addresses a broad range of topics and themes of interest to all who are involved with undergraduate mathematics. The volumes in this series are readable, informative, and useful, and help the mathematical community keep up with developments of importance to mathematics.

MAA Notes

11. Keys to Improved Instruction by Teaching Assistants and Part-Time Instructors, *Committee on Teaching Assistants and Part-Time Instructors, Bettye Anne Case,* Editor.
13. Reshaping College Mathematics, *Committee on the Undergraduate Program in Mathematics, Lynn A. Steen,* Editor.
14. Mathematical Writing, by *Donald E. Knuth, Tracy Larrabee, and Paul M. Roberts.*
16. Using Writing to Teach Mathematics, *Andrew Sterrett,* Editor.
17. Priming the Calculus Pump: Innovations and Resources, *Committee on Calculus Reform and the First Two Years,* a subcomittee of the Committee on the Undergraduate Program in Mathematics, *Thomas W. Tucker,* Editor.
18. Models for Undergraduate Research in Mathematics, *Lester Senechal,* Editor.
19. Visualization in Teaching and Learning Mathematics, *Committee on Computers in Mathematics Education, Steve Cunningham and Walter S. Zimmermann*, Editors.
20. The Laboratory Approach to Teaching Calculus, *L. Carl Leinbach et al.,* Editors.
21. Perspectives on Contemporary Statistics, *David C. Hoaglin and David S. Moore,* Editors.
22. Heeding the Call for Change: Suggestions for Curricular Action, *Lynn A. Steen,* Editor.
24. Symbolic Computation in Undergraduate Mathematics Education, *Zaven A. Karian,* Editor.
25. The Concept of Function: Aspects of Epistemology and Pedagogy, *Guershon Harel and Ed Dubinsky,* Editors.
26. Statistics for the Twenty-First Century, *Florence and Sheldon Gordon,* Editors.
27. Resources for Calculus Collection, Volume 1: Learning by Discovery: A Lab Manual for Calculus, *Anita E. Solow,* Editor.
28. Resources for Calculus Collection, Volume 2: Calculus Problems for a New Century, *Robert Fraga,* Editor.
29. Resources for Calculus Collection, Volume 3: Applications of Calculus, *Philip Straffin*, Editor.
30. Resources for Calculus Collection, Volume 4: Problems for Student Investigation, *Michael B. Jackson and John R. Ramsay*, Editors.
31. Resources for Calculus Collection, Volume 5: Readings for Calculus, *Underwood Dudley*, Editor.
32. Essays in Humanistic Mathematics, *Alvin White,* Editor.
33. Research Issues in Undergraduate Mathematics Learning: Preliminary Analyses and Results, *James J. Kaput and Ed Dubinsky,* Editors.
34. In Eves' Circles, *Joby Milo Anthony*, Editor.
35. You're the Professor, What Next? Ideas and Resources for Preparing College Teachers, *The Committee on Preparation for College Teaching, Bettye Anne Case,* Editor.

36. Preparing for a New Calculus: Conference Proceedings, *Anita E. Solow,* Editor.
37. A Practical Guide to Cooperative Learning in Collegiate Mathematics, *Nancy L. Hagelgans, Barbara E. Reynolds, SDS, Keith Schwingendorf, Draga Vidakovic, Ed Dubinsky, Mazen Shahin, G. Joseph Wimbish, Jr.*
38. Models That Work: Case Studies in Effective Undergraduate Mathematics Programs, *Alan C. Tucker,* Editor.
39. Calculus: The Dynamics of Change, *CUPM Subcommittee on Calculus Reform and the First Two Years, A. Wayne Roberts,* Editor.
40. Vita Mathematica: Historical Research and Integration with Teaching, *Ronald Calinger,* Editor.
41. Geometry Turned On: Dynamic Software in Learning, Teaching, and Research, *James R. King and Doris Schattschneider,* Editors.
42. Resources for Teaching Linear Algebra, *David Carlson, Charles R. Johnson, David C. Lay, A. Duane Porter, Ann E. Watkins, William Watkins,* Editors.
43. Student Assessment in Calculus: A Report of the NSF Working Group on Assessment in Calculus, *Alan Schoenfeld,* Editor.
44. Readings in Cooperative Learning for Undergraduate Mathematics, *Ed Dubinsky, David Mathews, and Barbara E. Reynolds,* Editors.
45. Confronting the Core Curriculum: Considering Change in the Undergraduate Mathematics Major, *John A. Dossey,* Editor.
46. Women in Mathematics: Scaling the Heights, *Deborah Nolan,* Editor.
47. Exemplary Programs in Introductory College Mathematics: Innovative Programs Using Technology, *Susan Lenker,* Editor.
48. Writing in the Teaching and Learning of Mathematics, *John Meier and Thomas Rishel.*
49. Assessment Practices in Undergraduate Mathematics, *Bonnie Gold,* Editor.
50. Revolutions in Differential Equations: Exploring ODEs with Modern Technology, *Michael J. Kallaher,* Editor.
51. Using History to Teach Mathematics: An International Perspective, *Victor J. Katz,* Editor.
52. Teaching Statistics: Resources for Undergraduate Instructors, *Thomas L. Moore,* Editor.
53. Geometry at Work: Papers in Applied Geometry, *Catherine A. Gorini,* Editor.
54. Teaching First: A Guide for New Mathematicians, *Thomas W. Rishel.*
55. Cooperative Learning in Undergraduate Mathematics: Issues That Matter and Strategies That Work, *Elizabeth C. Rogers, Barbara E. Reynolds, Neil A. Davidson, and Anthony D. Thomas,* Editors.
56. Changing Calculus: A Report on Evaluation Efforts and National Impact from 1988 to 1998, *Susan L. Ganter.*
57. Learning to Teach and Teaching to Learn Mathematics: Resources for Professional Development, *Matthew Delong and Dale Winter.*
58. Fractals, Graphics, and Mathematics Education, *Benoit Mandelbrot and Michael Frame*, Editors.
59. Linear Algebra Gems: Assets for Undergraduate Mathematics, *David Carlson, Charles R. Johnson, David C. Lay, and A. Duane Porter,* Editors.
60. Innovations in Teaching Abstract Algebra, *Allen C. Hibbard and Ellen J. Maycock*, Editors.
61. Changing Core Mathematics, *Chris Arney and Donald Small,* Editors.

MAA Service Center
P. O. Box 91112
Washington, DC 20090-1112
800-331-1622 fax: 301-206-9789

Foreword

"Hence we must believe that all the sciences are so interconnected, that it is much easier to study them all together than to isolate one from all the others. If, therefore, anyone wishes to search out the truth of things in serious earnest, he ought not select one special science, for all the sciences are cojoined with each other and interdependent." — Descartes

This volume discusses how core mathematics (the first two years of instruction) should change over the next five to ten years. The perspectives considered in this analysis include the goals and contents of the courses, the anticipated advances in technology, the development of a more interdisciplinary academic culture, and the instructional techniques associated with teaching core mathematics courses. The genesis of this volume was an interdisciplinary workshop held at the United States Military Academy (USMA) where mathematicians, engineers, and physical scientists analyzed the future role of undergraduate mathematics.

The editors sincerely thank the workshop participants for their active engagement in this process and the authors of the enclosed papers for their contributions. We want to recognize the extra contributions made by Gary Krahn, Joseph Myers, Patrick Driscoll, and Kathleen Snook, as editors of the papers in Part 2 of the volume. We also thank the sponsors of the workshop: National Science Foundation (through Project INTERMATH, part of the Mathematics Across the Curriculum (MATC) initiative) and the Mathematical Association of America (through the Committee on the Undergraduate Program in Mathematics (CUPM) and its subcommittees Calculus Reform and the First Two Years (CRAFTY) and Mathematics Across Disciplines (MAD)). A special thanks is also given to Dr. William Wulf, President of the National Academy of Engineering, for providing an inspiring keynote address at the workshop on the "Urgency of Engineering Education Reform."

The workshop participants (by discipline) were as follows:

Engineers:

Jim Dally	University of Maryland (emeritus)
Jeff Froyd	Texas A&M University
Mary Goodwin	Iowa State University
Jack Grubbs	Tulane University
Mike McGinnis	United States Military Academy
Richard Plumb	SUNY Binghamton
Steve Ressler	United States Military Academy
Andre Sayles	United States Military Academy
John Scharf	Carroll College
Bob Soutas-Little	Michigan State University
Bill Vanbuskirk	New Jersey Institute of Technology
Bill Wulf	President, National Academy of Engineering

Physicists:

Bob Fuller	University of Nebraska (visiting USMA)
Heidi Mauk	United States Air Force Academy
Tom Lainis	United States Military Academy
Jim Stith	American Institute of Physics

Mathematicians:

Don Albers	Mathematical Association of America
Chris Arney	United States Military Academy
Bill Barker	Bowdoin College (CRAFTY)
Lida Barrett	United States Military Academy (emeritus)
Tom Berger	Colby College (CUPM)
Lisette dePillis	Harvey Mudd College
Ray Cannon	Baylor University (visiting USMA)
John Dossey	Illinois State University
Penny Dunham	Muhlenberg College
Laurette Foster	Prairie View A&M University
Frank Giordano	COMAP
Bill Haver	Virginia Commonwealth University
Gary Krahn	United States Military Academy
James Lightbourne	National Science Foundation
Dave Lomen	University of Arizona
Joe Myers	United States Military Academy
Shirley Pomeranz	University of Tulsa
Fred Rickey	United States Military Academy
Don Small	United States Military Academy
Kathi Snook	United States Military Academy
Elizabeth Teles	National Science Foundation
Frank Wattenberg	Texas Instruments
Brian Winkel	United States Military Academy
Lee Zia	National Science Foundation
Paul Zorn	St. Olaf College

In this volume, the authors identify issues and make recommendations for future course design. The volume details the interdisciplinary needs of our students and provides a basis for the construction of the core mathematics curriculum for the first one or two years of college. The contents of the volume include in Part 1 (Overview): historical development of core mathematics and its teaching, future considerations, description of an integrated program, description of a proposed inquiry and modeling program, example of such a program, and description of an environment for change. Part 2 (Commentary) contains 21 articles written by curricular experts who attended the workshop. The Appendices contain three example Interdisciplinary Lively Application Projects (ILAPs) that show the type of problems that core students are expected to confront and "solve".

Contents

Glossary of Terms

ABET	Accreditation Board of Engineering and Technology
ALEKS	Assessment and LEarning in Knowledge Spaces
AMATYC	American Mathematical Association of Two-Year Colleges
AMS	American Mathematical Society
ATLAST	Augment the Teaching of Linear Algebra through the use of Software Tools
CAS	Computer algebra system
CBMS	Conference Board of the Mathematical Sciences
CODEE	Consortium for Ordinary Differential Equations Education
COMAP	Consortium for Mathematics and Its Applications
CRAFTY	Calculus Reform and the First Two Years (subcommittee of CUPM)
CUPM	Committee on the Undergraduate Program in Mathematics (MAA)
DDS	Discrete dynamical systems
DUE	Division of Undergraduate Education of the NSF
EDPIC	Problem solving using Exploration-Description-Planning-Implementation-Checking
FCI	Force Concept Inventory
IFYCSEM	Integrated First-Year Curriculum in Science, Engineering, and Mathematics at Rose-Hulman Institute of Technology
ILAP	Interdisciplinary lively application projects
INTERMATH	NSF-sponsored consortium centered at COMAP and USMA
MAA	Mathematical Association of America
MAD	Mathematics Across Disciplines (subcommittee of CUPM)
MATC	Mathematics Across the Curriculum
ME	Mechanical engineering
NCTM	National Council of Teachers of Mathematics
NSF	National Science Foundation
ODE	Ordinary differential equations
PSL	Problem solving laboratory
PUFM	Profound understanding of fundamental mathematics
STEM	Science-Technology-Engineering-Mathematics
USMA	United States Military Academy

Introduction

How should core mathematics change over the next five to ten years? We investigate this question and provide answers by developing historical and future perspectives, identifying major issues, recommending approaches to achieve program objectives, and making suggestions for future course design. The needs of our students in preparation for their service to society provide the basis for the construction of the core mathematics curriculum (required base-line service courses) for the first one or two years of college.

Foremost among our ideas on core mathematics is the following principle: "All students (should) have access to supportive, excellent undergraduate education in science, mathematics, engineering, and technology, and all students (should) learn these subjects by direct experience with the methods and processes of inquiry."[1] William Wulf, President of the National Academy of Engineering, provided us with additional issues, guidelines, and standards for core education. He addressed a multi-disciplinary group working on future curricular development with his presentation entitled "The Urgency of Engineering Education Reform." His remarks on engineering reflect the situation in mathematics in many ways. It's from several of his points, some that we provide here, that we draw motivation and inspiration for many of the ideas presented for changing core mathematics. Wulf stated as his major point: "I think we ought to be seeing a watershed change in engineering education—it is not happening. I am very impatient about it. … I fully appreciate that if you go to any engineering school you are likely to find some innovative things happening. What is not happening is the center of gravity moving in any substantive way."[2] In this volume, we propose a shift in the center of gravity in core mathematics. We recommend a change in focus. The major focus of core mathematics should be on process (problem solving and thinking — modeling and inquiry), not in content (facts and techniques). Also, the content portion of core mathematics should be relevant to society and science and must be modernized for future utility in the Information Age.

Wulf's discussion of the forces changing engineering also relates to our society's needs for change in mathematics. In today's information-based society, quantifying and processing data is required in many professional fields and life experiences. This "mathematization" of society definitely has an impact on core mathematics, just as complexity is changing the future of engineering. As Wulf stated: "The complexities of the design space and the constraint set are exploding, and there are relevant social changes — the expanding role of engineers in industry and the globalization of engineering."[3] Wulf went on to explain the increased complexity of our world, as he compared the construction materials available to two generations of engineers,

---

[1] Mel George, President emeritus of St. Olaf Collage; author of NSF report, "Shaping the Future: New Expectations for Undergraduate Education in Science, Mathematics, Engineering, and Technology," p. 11, *Towards Excellence, Leading a Doctoral Matheamtics Department in the 21st Century,* American Mathematical Society Task Force on Excellence, John Ewing, Editor, 1999.

[2] William Wulf, "The Urgency of Engineering Education," p. 234, *Proceedings of the Interdisciplinary Workshop on Core Matheamtics*, West Point Press, 2000.

[3] Ibid., p. 235.

his own with his father's: "My father was an engineer. ... For my father there was a little book on a shelf, a little thin book, of the materials that he had as an option to design with. There were a half a dozen different kinds of steel, there were a few kinds of bronze, plastic was not in his vocabulary, fibers were not in his vocabulary, composite materials were not something he considered. Now we are talking about designer materials. That is the ability for an engineer to say these are the properties that I want the material to have and at least potentially the possibility of producing that material for that subject. Literally that thin book has become an infinite set of options."[4] Similarly, the kinds of mathematics available to undergraduates are broader and more varied than ever before. We believe students need to experience the fundamental processes of mathematics as empowering them to understand the complexity and technical advances of the changing world, not just experience a deeper study in a rich, but narrow topic in mathematics.

The changes advocated and discussed in this volume are not restricted to the mathematics curriculum. Other factors, such as pedagogy and student growth, are closely related to the changes being advocated and are, therefore, discussed as well. Such changes are so fundamental that the academic culture itself must also evolve. Again, William Wulf's presentation provides a backdrop for the authors' discussion of these topics: "It seems to me that this really underscores the fact that the engineer who is trained superbly in a technical sense but does not understand the cultural and social issues in a very broad sense, in a multicultural way, is really useless. ... The pace of change is itself a change. ... I think the important point is that it has not been part of the engineers' culture to feel responsible for their own lifelong learning and I think that has to change. ... [There are many things] that need to change, curriculum, pedagogy, faculty award system, the need for formalized lifelong learning, preparation for K through 12, and technological literacy of the general population."[5] We've tried to address some of these issues as they relate to core mathematics. Indeed, to do all of this, we agree with Wulf, the center of gravity must shift.

We've arranged the volume in two major Parts. Part 1 contains a historical presentation of course development and pedagogical change, the philosophy and components of a future integrated/interdisciplinary core program, a framework for a proposed inquiry/modeling core program, an example of such a program, and a description of an environment where progress and change are integral elements of the academic culture. While some of these ideas are novel in their approach by emphasizing process over content, we believe that there are many advantages to our suggestions as we seek to meet society's needs as it enters the Information Age. Part 2 contains 21 articles written by curricular experts from several disciplines. These authors are engaged in curriculum issues and attended an interdisciplinary workshop focused on the future of core mathematics. The authors are grouped into four categories based on their perspective taken on this curricular question. The four major perspectives considered in this analysis of the curriculum include: 1) the goals and contents of courses and programs, 2) the anticipated advances in technology, 3) the development of a more interdisciplinary academic culture, and 4) the instructional techniques associated with teaching core mathematics courses. The articles provide insights into several important issues, to include the skills to be attained, the problems that ought to be solvable after two years of study, the nature of assessment tools that should be employed, and the "learning environments" necessary for student growth.

The types of questions analyzed and discussed in the papers through the four perspectives include:

Interdisciplinary Culture

What is the impact of mathematics reform on the partner disciplines?
How should science education reforms affect mathematics instruction? (and vice versa)
How is mathematics effectively integrated into the undergraduate curriculum?
When should calculus be taught and what other courses are needed?
How is on-going involvement of the partner disciplines maintained?

---

[4] Ibid.

[5] Ibid., p. 238.

Technology

How should technology affect what and how we teach?
What are the strengths/drawbacks of different technology choices?
How should we match technology choices to the intended audience?
What are the possible effects of future technology?

Goals and Content

What are the important and difficult content choices?
How should we balance theoretical understanding with computational skills?
How should we match goals to the intended audience?
How do modeling and applications fit into the curriculum?
What are "high standards" and how can they be achieved?
When should calculus be taught and what other courses are needed?

Instructional Techniques

What are the strengths/drawbacks of different instructional methods?
How should we choose and integrate various instructional methods?
What methods best increase success of underrepresented groups?
How do the learning media affect reading, writing, and problem solving?
How should we build theoretical understanding?
How should we align the "achieved curriculum" and the "intended curriculum"?
What guiding principles arise from educational research?
Should calculus be a laboratory (discovery) course?
How should we assimilate the skills-proficient high school students?

The Appendices provide three Interdisciplinary Lively Application Projects (ILAPs) that are examples of the types of problems that we feel should be assigned to students in core mathematics programs to develop both their mathematical talents and their interdisciplinary perspective.

Part 1

Overview

Historical Development of Core Mathematics and Its Teaching

Historic Tour of Undergraduate Mathematics

The advancement and perfection of mathematics are intimately connected with the prosperity of the State.
—Napoleon

Throughout history, mathematics has served as a powerful tool for the civilized world. There are many examples where mathematics has contributed significantly to the advancement of society. While we can establish a strong connection between mathematics and progress in society, for this presentation we are interested in two questions, "What mathematics should we teach our undergraduates?" and "How do we teach mathematics to our undergraduate students?" We discuss those questions through a combined disciplinary and historical perspective of undergraduate mathematics education in America.

Benjamin Franklin highlighted the utility of mathematics in his 1738 paper entitled "On the Usefulness of Mathematics" when he wrote: "What science can be more noble, more excellent, more useful for men, more admirably high and demonstrative, than that of mathematics?" However, in Franklin's day, mathematics was frequently taught as an art, exercising the mind in reasoning, memorizing, analyzing patterns, and reciting formulas, proofs, and theorems. Of course, the practical, professional side of mathematics also empowered business people, farmers, surveyors, and navigators of colonial America. Some of these practical skills were found in college-level academic courses (e.g., geometry for surveyors). It was important for the growth of our nation that the colleges of early America taught mathematics as a professional tool as well as an art.

In the early nineteenth century, undergraduate mathematics began to be taught as the language of science. This kind of mathematics has structure, process, and utility for communicating science. New subjects in mathematics and science slowly were added to curricula, teaching styles were refined (applied problems were solved and analyzed), and college graduates who knew the language of science were being produced. These graduates could assemble mathematics to analyze technical problems and some used their mathematics and science to perform engineering, building better roads, buildings, bridges, canals, and railroads. Many American colleges developed curricula that taught mathematics as the language of science while preserving many of the aspects of teaching mathematics as an art. America's education system began to produce highly skilled college graduates, some becoming engineers and technologists. Mathematics empowered college graduates to become productive, successful citizens. This period is called "the initial phase of *Mathematisation* (sic) in America, on the grounds that Americans before the early nineteenth century were not a 'calculating people.' … Yet by 1900 the US was already well on the way to becoming the number-obsessed culture it has remained down to today." [10, 19]

By the beginning of the twentieth century, high school and college students had the opportunity to learn mathematics as a language of science and utilize mathematics to solve problems. Society had advanced in its development and use of technology (e.g., steamships, trains, automobiles, telegraphs, electric lights). What part would mathematics play in the new era? Fortunately a new role for undergraduate mathematics was available. In addition to being an art and a language, college-level mathematics had become a science — the science of measurement. As a science, mathematics offered a systematic method to solve specific problems, i.e., those that involved measurement. By the early twentieth century, the science component of mathematics was available to undergraduates. Many peoples' jobs required them to perform quantitative measurements and quantitative decision making. Our society had embraced technology. New devices like cars, planes, radios, refrigerators, televisions, and telephones were developed and refined using the science of mathematics. Almost all citizens, especially college graduates, used technology in their personal and professional lives. College graduates trained as managers and professionals were expected to understand, design, and optimize sophisticated plans and operations; use and maintain complex mechanical and electrical equipment and transportation systems; design and implement efficient schedules; and understand entirely new technological devices. Mathematics requirements for undergraduates had evolved to highlight the understanding and use of mathematics as a science.

A Look to the Future

What challenges does the twenty-first century and the dawning of the information age bring to undergraduate mathematics? Modern technologies in the forms of calculators, computers, and information networks are tremendous tools for communication, visualization, and problem solving. The art, language, and science roles of mathematics continue to change dramatically. Technological tools actually perform much of what was considered undergraduate mathematics in the not-so-distant past. People now face new technological and quantitative challenges. College-educated managers and professionals are required to process data and synthesize information, use and understand information technology, optimize elaborate plans, confront complexity, and leverage new technologies. The diverse missions of today's businesses and industries require people with a multitude of skills to confront the myriad of challenges of the modern world. Schools' educational goals are beginning to take these new challenges into account. Today's college core mathematics programs need to produce creative, confident, competent problem solvers. Modeling (forming and analyzing problems, using technical tools, and implementing solutions) with an emphasis on interdisciplinary problem solving (working in teams) becomes an essential component of modern undergraduate mathematics. Less emphasis is needed in teaching the skills and techniques best performed by technology. College graduates will need to learn how to use newly developed technological tools to solve problems from every facet of life (physical sciences, life sciences, social sciences, behavioral sciences, political sciences, technology, and humanities). They will need to become quantitative and interdisciplinary problem-solvers to serve society and satisfy the world's diverse needs.

Over these past two centuries the needs of the society have changed and with it the undergraduate mathematics curriculum. College graduates, in filling their role as citizens, business leaders, and productive professionals, provide the intellectual and technical leadership of society and are required to be professionally competent. More than just knowing one's job or establishing one's proficiency, competence in the twenty-first century will include having the attributes of creativity and confidence, skills in quantitative problem solving, and the ability to learn. Mathematics as an art, language, science, and problem-solving tool, contributes greatly to the development of these essential attributes. More than ever, our students need to study mathematics because of its importance in the everyday world. Many daily situations bring people into contact with mathematics, including buying products, conducting business, banking, producing products, managing people and technology, communicating, and using science and technology. Our nation's undergraduates must learn core-level mathematics to efficiently and effectively live in and contribute to the technological world of the twenty-first century. Our core mathematics programs must keep pace with the rapid changes that are and will continue to take place.

Historical View of Pedagogical Developments and Teaching Tools

But of all the sciences cultivated by mankind, none are more useful than mathematics, to call forth a spirit of enterprise and enquiry. — Consider Sterry

The mathematics classroom and its associated teaching tools have changed over the history of core mathematics in America. Lectures have always been the mainstay to explain concepts and transfer information. The primary teaching tools in mathematics classrooms, the blackboard and chalk, were first used by teachers around 1800. This teaching tool provides excellent opportunities for visualization and interaction, which are important in showing students the necessary techniques and skills of mathematics. Overhead projectors, which enhanced these capabilities, were utilized first during the middle of the twentieth century. These tools still serve a role in the mathematics classroom of today.

The familiar textbooks used by American students that portray in one volume the theory, concepts, and motivation for many of the topics in the subject, along with providing numerous examples and exercises, were also developed in the early nineteenth century. A classic example of the growth of this teaching tool is the modern calculus textbook with over 1000 multi-colored pages and hundreds of figures, examples, and exercises. Student notebooks and portfolios have also been around in various forms for over 200 years, while the amount of detail and kinds of information included have changed. Teaching technology in the form of visual displays and physical models have supplemented classroom instruction and textbook presentation.

Computation, a key element of mathematical instruction, has undergone the most significant changes. These changes have had a significant impact on pedagogy in mathematics. Computation was first performed in colonial and early America by numeric tables and slide rules, later by mechanical calculators and then electronic calculators. The late twentieth century brought mathematics two all-purpose teaching tools, the computer — especially the personal computer, and the graphing calculator. Both these devices, which can perform many of the tasks provided by various teaching technologies in the past, are significantly impacting pedagogy. The computer or graphing calculator in the classroom has the potential to provide increased visualization, textual material, interaction, information, and computation. Over the last decade, the tremendous capabilities of computer algebra systems found on both computers and calculators have dramatically affected both what mathematics is taught and how it is taught to undergraduates. Today, these two devices seem to be merging into one. Calculators become more powerful with more capabilities. Computers become smaller and more portable. The development of the computer-calculator is revolutionizing the way undergraduate core mathematics is taught..

Today there is considerable support for the learning theory of constructivism in mathematics teaching and learning. Many programs and courses also have as a goal the development of students as life-long learners. In order to accommodate constructivism and the development of life-long learning skills, active learning or student-centered teaching methods are more prevalent. In addition to traditional reading, lecturing, and testing, several new methods and issues are being used and considered in the modern undergraduate core program: classroom questioning and discussion, (applied) problem solving and modeling, technology, explorations and discovery, multiple representations of mathematics, writing, various types of assessment instruments, smaller section sizes, and collaborative or cooperative groups.

A first step toward a more interactive classroom is to have students enter the mathematics discourse. Many of today's instructors promote questioning and discussion in the classroom. Since students are also interested in the relevance and worth of the mathematics they are studying, problem-solving activities, especially realistic applications, are useful motivators. Applications provide students a picture of how the mathematics they are learning is used and connected to subjects in other disciplines. Many modern core mathematics programs are technology dependent, incorporating calculators and/or computer algebra systems. In laboratory classrooms, technology is usually integral to the course. Discovery and exploration require the proper technological tools for testing conjectures and generalizing theories. Computer Algebra

Systems and calculator-based laboratory devices are valuable components for courses with goals to develop exploration and discovery. The "rule of four" was introduced based on the belief that multiple and integrated perspectives — numerical, graphical, analytical, and word representations of mathematics — can best develop conceptual understanding. More than just a representation of mathematics, student writing both enhances and reveals students' understanding.

As we change how and what we teach, we continue to change how and what we assess. With the advent of technology, a proficient calculator user can pass many traditional exams without understanding the material. Assessment plans that use a variety of problem types and different modes of presentation provide a more comprehensive view of students' understanding. Cooperative and collaborative groups can also assist in more interactions between teachers and students. Group work facilitates students discussing mathematics in their own words. They verbalize and explain their reasoning to peers. This, in turn, organizes their conceptions for improved and deeper understanding. Additionally, group and teamwork experiences are becoming essential for future study and successful employment.

The pedagogy and teaching tools of the modern core mathematics class are different than those of traditional classes in earlier times. Technology has, and will continue to have a tremendous impact on the way we teach.

Overview of Calculus Reform

The calculus course is at the same time a culmination and a beginning. It is a place where many of the ideas and techniques learned in the secondary mathematics curriculum are pulled together, the place where many of the naturally occurring questions from those courses can be answered in a satisfying way. But it also is the foundation for the study of the natural sciences, engineering, economics, and an ever-increasing number of the social sciences.

— A. Wayne Roberts, MAA Notes 39, 1996

Because many core mathematics programs currently contain calculus and the calculus courses at many schools are undergoing reformation, we give a short development and status of America's calculus reform movement. This reform movement is often pegged to the Tulane Conference in 1986, but it was rooted in the problems of the 1950s and 1960s when calculus became a first-year course. At Tulane, the conference participants identified five problems of typical calculus courses and put together a blueprint for action and improvement. The five problems were:

- Too few students successfully complete calculus
- Students mindlessly manipulate symbols without conceptual understanding
- Faculty are frustrated over student motivation and preparation
- Calculus acts as an unnecessary filter for many disciplines
- Instead of endorsing appropriate use of technology, faculty often prohibit its use

The blueprint contained three mandates:

- Focus on conceptual understanding that uses multiple representations (graphic, numeric, and word, as well as symbolic) and gear the course to average students
- Teach with various modes of instruction and use technology to engage the students
- Foster an inclusive spirit in the reform with an emphasis on cooperation and broad dissemination

After the initial discussions, organizations such as the National Science Foundation and the Mathematical Association of America began to support reform efforts and initiatives. Several major reform projects and many smaller, grass-root projects were started in the late 1980s. After some refinements and programs were tested at various schools, the following statement made in 1989 was one of the first summaries of the calculus-reform movement: "In spite of the innovative nature of the proposals, the contemplated changes are not in the overall content of the course, but rather in the expectation of better student understanding;

in greater stress on applications and links with other disciplines; in the utilization of numerical methods and computer techniques; and in encouraging a fresh approach to teaching." [54] In addition to developing new textual materials and technology manuals, faculty development workshops fueled the engine of reform and spread reform ideas and resources around the country. The calculus reform movement became a grass-root effort involving all sectors of the mathematics community (teachers and researchers, two- and four-year schools, liberal arts and technical schools) and directed towards improving instruction for all students. A more recent assessment of calculus reform was given in *The Chronicle of Higher Education* in 2000: "It swept college campuses during the early 1990s in response to educators' concerns that students knew how to use techniques to solve formulas but didn't understand what the formulas were for. Although universities did change course content—story problems were introduced, and less importance was given to memorizing techniques and applying them repeatedly—the biggest changes were pedagogical. Universities traded large lecture courses for smaller sections of calculus, and students were asked to give oral presentations and write papers rather than manipulate formulas over and over." [58]

The current status of calculus reform shows the following results:

- Calculus reform is widespread (most students are in courses using some of the elements of calculus reform and nearly all calculus textbooks incorporate some aspects of reform).
- More emphasis is placed on conceptual understanding rather than theoretical understanding (fewer proofs and more non-standard problems).
- More balanced representation (graphic, numeric, symbolic, and words are integral parts of courses) is utilized.
- More emphasis on authentic applications (interdisciplinary, real data) is offered.
- Significant changes in pedagogy (projects, writing, cooperative groups, variety of assessments) are present.
- Technology is entrenched and empowering many curricular and pedagogical changes (calculators, computers, CAS, laboratory/discovery goals).
- Strengthened connections with partner disciplines are in place.

Similarly, there are issues and challenges still to be confronted and overcome:

- Reform activities are superimposed not completely integrated into many programs
- Questions remain on the proper role of technology (in learning and performing skills)
- Proper assessment is still difficult
- Research questions on student learning and thinking are not resolved
- Few programs have been developed with substantially revised content and focus

Considerable progress has been made in improving calculus. Concerns over fundamental skills still linger. What should students be able to do without a calculator? What are the importance of hand calculations and skills? Which concepts/skills should be learned first by hand, but once understood shifted to a calculator procedure? The mathematics community must understand and resolve these issues as we prepare to educate these students in the twenty-first century.

The Impact of Science on Mathematics and Teaching Mathematics*

> *New sciences are emerging. And they measure themselves not by any of today's scientific yardsticks, but by the needs of tomorrow's technology.* — Keith Devlin [14]

Educational programs must evolve to meet the future needs of society. However, this does not always occur easily. Change is difficult, and as the pace of change increases, our academic programs must adapt

* This portion of the paper is adapted with permission from: Arney, D. C., "Education for the Future: Mathematics for Understanding the New Sciences, PRIMUS, vol. VIII, Sept. 1998, pp. 240–252.

and adopt necessary changes to serve society. In general, academic institutions adopt innovations slowly, sometimes retarding the growth of students. [42] Most educational crises and reform movements address problems that have arisen many times before or have been present for an extended period. The calculus reform movement of the 1990s had its roots in the problems of the 1950s and 1960s. The reform movement, while substantially changing pedagogy, still left calculus as the cornerstone of most core programs. There was a remaking of an existing course, but little change in course content. A question to consider is how and when new topics and new courses should enter the curriculum. We discuss here undergraduate core mathematics courses and suggest that new core courses emphasizing the processes of modeling and inquiry and the content of discrete dynamical systems, are needed to prepare students for success in the future era of the information age where new sciences and technologies like complexity theory, information science, and genetic coding, and new teaching tools like the internet, will be prevalent.

Just as an understanding of differential equations is necessary to solve many problems involving continuous rates of change and their relations, as we found in various engineering problems, the analogous situations in the new sciences need the mathematics of changing discrete phenomena, in particular, discrete dynamical systems. While this subject has previously been thought of as an elective, post-calculus course for specialists, we recently have seen a discrete version of this course enter the core as a required first-year course at several schools. Similarly, modeling and inquiry were delayed to higher-level course work. Many students never took courses that embedded content with problem solving and thinking. It is time for that to change.

To set the stage for this discussion, we present a history of the debut of some core undergraduate mathematics courses, give information about the new sciences (complexity theory, information science, and genetics), describe a freshman-level discrete dynamical modeling course being implemented at some schools, and outline the components of a proposed course in modeling and inquiry which could utilize not only dynamical systems, but also concepts from calculus, matrix algebra, differential equations, and probability and statistics.

History of Core Mathematics Courses

As is well known, physics became a science only after the invention of differential calculus.
— Bernhard Riemann (1882).

Geometry. If we look back to colonial America, we find that college-level mathematics was not generally taught for application, but for exercising the mind. At the handful of colleges in existence in the eighteenth century, mathematics was either not required or its application was ignored. The one exception was geometry, which was taught for both exercising the mind (proofs and constructions) at most colleges and for its application at a few schools to a very few students. The application of geometry was needed for surveying and navigating and, therefore, was taught as a prerequisite course for those professions.

Descriptive Geometry. At about the time of the American Revolutionary War, civil engineering (a new science) and its applications of designing bridges, canals, roadways, and fortifications needed more than basic geometry. As a result, a new course, descriptive geometry, entered the curriculum. Since not everyone was going to be a civil engineer, descriptive geometry was not required for all students.

College Algebra and Trigonometry. While new mathematics was available for inclusion in core programs during the rest of the nineteenth century, only a few new topics entered into college core programs. One could argue that no new science was developed sufficiently to warrant such an action or that mathematics education programs were slow to develop their supporting courses. Therefore, many schools required the traditional mathematical thinking courses with algebra and trigonometry as the core mathematics requirement and have maintained that requirement ever since.

Calculus. The first school in America where all students were expected to study engineering was the United States Military Academy (USMA). Shortly after its founding in 1802, students took algebra, geometry, trigonometry, and descriptive geometry with the intent to use these mathematical skills in their professions. As engineering became more sophisticated and machinery was being designed by mechanical engineers using concepts from physics, calculus was required for understanding this new science. This new required core course (calculus) was added to the curriculum of many emerging engineering- or science-based schools in the United States by the middle of the 19th century. Since then, the number of disciplines requiring calculus has increased. Now, many programs besides engineering require calculus, and it is a core course in many college programs.

Probability and Statistics. While statistics courses had been available in many programs in the late nineteenth century, it was not a general requirement or core course. Operations research, which blossomed during World War II, was the next new science affecting the mathematics curriculum. To use and understand many tools in operations research, courses in probability and statistics (with topics like least squares) and expanded matrix algebra topics were introduced into the core program.

Discrete Mathematics. The war years also produced the computer, and the discipline of computer science was born. The requisite mathematical topics (propositional and Boolean logic, algorithms, combinatorics, sets, trees, graphs, induction, and networks) were assembled into one course [17,34]. The new science (computer science) spawned another new core mathematics course (discrete mathematics).

Discrete Dynamical Systems. Some programs see the new science of chaos generating the need for a new core course in dynamical systems. The development of chaos theory and its associated study of nonlinear dynamics and discrete modeling in the 1970s and 1980s are giving rise to core courses in discrete dynamical modeling during the 1990s. [38]

Integrated Programs. A recent phenomenon in education is integrated programs that combine several topics (courses) into one thematic or topics course. High school programs in some states have contained integrated courses for several decades. Several college programs followed suit in the 1980s and 1990s. The topics course in discrete mathematics discussed above is an example. Other programs have integrated topics around a theme (rates of change: derivatives, differential equations, and difference equations) (accumulations: integrals and series).

Future Considerations

What will be next?

The whole of mathematics consists in the organization of a series of aids to the imagination in the process of reasoning. — Alfred North Whitehead

This historical development does not mean to imply that all or even most college core mathematics programs added all these new courses or should blindly follow the model of new science creating new courses. Moreover, we have not mentioned when and how courses leave the curriculum or how they evolve over time. Currently courses containing geometry, algebra, trigonometry, some discrete mathematics, some probability and statistics, and some pre-calculus (functions) are required in many high school programs. This discussion assumes that undergraduate students have had success in these courses. There has been considerable discussion over the past decade about the content of the mathematics curriculum, especially the make-up of the undergraduate core courses. [15,49,51] Looking at this history, there seems to be a connection between the development of new science, the availability of new technologies, and new mathematics courses entering the curriculum at the core level. Given the important and fundamental role of mathematics in science and technology in mathematics, these connections make sense. Curriculum changes should be informed by the content of other courses, the availability of teaching tools, and ultimately the needs of society.

Technology in the form of personal computers, internet access, and calculators have empowered students with unprecedented computational and visualization skills, tremendous information access, and sophisticated exploratory and discovery capabilities. Merely teaching students the content and skills that their technology already provides is not enough for a modern core program. We must teach our students the appropriate use of these powerful tools and the content that is most relevant to their future needs as successful students, productive graduates, and informed citizens. The real power of mathematics for undergraduates is in its processes of thinking and inquiry. These processes are critical in problem solving and decision making.

Information Science, Complexity, Genetics

[I] direct my thoughts in an orderly way; beginning with the simplest objects, those most apt to be known, and ascending little by little, in steps as it were, to the knowledge of the most complex; and establishing an order in thought even when the objects had no natural priority one to another. — Rene Descartes, *Discours de la Methode* (1637)

It has been mentioned that the new science of chaos is having an impact upon the core mathematics program. Several schools now include dynamic systems courses in their core programs [7,14,37]. Let's look a bit deeper into the current situation of science as we leave the industrial age and enter the information age.

Through computer processing, we now have machines with access to abundant amounts of data. We easily acquire more and more data. However, we have the problem of transforming this data into usable information. In many cases, we do not have the ability to represent it, analyze it, use it, process it, understand it, model it, or improve society with it. Having data locked away in a machine does not mean you have information. This is why a new science is developing, and this collective enterprise currently is called information science. In many ways, information science is about having machines transform data into information (or intelligence) as the human mind does. We need machines to perform rudimentary thinking processes. Therefore, to develop information science, we need to understand the complexity of both the human mind and machines holding data. One of the basic means of modeling the transformation processes of data analysis is through dynamical systems.

It has been discovered that many processes and systems in diverse topics such as weather, ecology, biology, engineering, economics, medicine, politics, and warfare, while seemingly unpredictable, follow deterministic models. These processes demonstrate behaviors and structures that vary from linear to nonlinear, predictable to random, and discrete to continuous. In reality, very few of the useful systems are simple, linear, or completely deterministic or random. Many phenomena are complex in their behavior: sensitive to slight changes in conditions; quite erratic or aperiodic; yet bounded. Such behavior is studied in chaos theory. From the abstraction of chaos to other areas of science, many new terms and concepts have arisen: fractals, nonlinear dynamics, cellular automata, computability, artificial life, intelligent systems, knowledge engineering, natural language, learning systems, neural networks, and complexity. In its broadest sense, complexity theory is the collective enterprise of understanding the structure, measure, and classification of these phenomena [47]. Chaos could be considered part of this new science. This science discerns that physical, biological, cognitive, and, in general, information systems are complex and need a new structure to understand their complexity. Once again, dynamical systems provide a way to model these phenomena.

The role and impact of information science on our society can be seen in the development of new technologies. Over the past 40 years, scientists have discovered and modeled the properties and structure of DNA. Recently, the information in DNA, chromosomes, and genes (the genetic code) has also been determined. In the future, the mathematical models of the structure and role of DNA will make gene therapy and genetic engineering accessible. As the human genome is classified over the next decade or so, important decisions about this technology must be made by society. We must equip our citizens and decision makers with the mathematical and scientific means to understand this technology and its consequences.

> *The fact is you can have intelligence imbedded in everything. There will not be a product produced 20 years from now which doesn't have some degree of intelligence ... Information technology is going to be in everything that engineers produce. And discrete mathematics, not continuous mathematics, is the underpinning of information technology. Biology and chemistry are going to become as fundamental as continuous mathematics and physics.*
>
> — William Wulf, President of the National Academy of Engineering (November 1999)

What are the connections between information science, complexity theory, and genetic coding? In order to develop information science and utilize genetics, we need to understand physical and cognitive complexity to process data and codes into information. These closely related sciences will provide society its greatest opportunities and challenges as we enter the information age. We need to prepare our students and our society to deal with these challenges and make wise decisions. The mathematical foundations for these sciences, dynamical systems and modeling and inquiry, should be taught in our core mathematics programs. Not only do our future scientists need this background, but also our managers, leaders, technologists, decision makers, and informed citizens need it as well. Complexity theory, information science, genetics, and their technical applications will greatly influence our world during the 21st century. [20,21,23,24,27] Our core educational programs must reflect this requirement.

Discrete Dynamical Modeling

The mind is not a vessel to be filled, but a flame to be kindled. — Plutarch

The basic concept in discrete dynamical modeling is that the future is predicted by understanding the present and adding to it the hypothesized change over the interval of interest. Discrete dynamical models (difference equations), with sufficient initial data, are always solvable by iteration. The prerequisite mathematics to learn and perform elementary discrete dynamical modeling is algebra. Therefore, this topic is accessible for first-year students without an investment in learning the more sophisticated calculus concepts needed to study continuous dynamics (differential equations). Many topics, especially modeling, inquiring, reasoning, and computing, that are traditionally covered in higher-level courses are accessible to freshmen taking an introductory discrete dynamical modeling course. Through discrete dynamical modeling the foundations of our new sciences are available to all students at the core level.

Goals: A valuable set of goals for a core mathematics program might include: students acquiring important and fundamental processes for future application; students developing sound, logical thought processes relevant to future science; students learning the fundamentals of data structures and processing; students developing an awareness and appreciation for interdisciplinary perspectives in solving problems; and students learning how to learn. By achieving these goals, successful students could formulate intelligent questions, reason and research solutions using new scientific principles, and be confident and independent in their future work. A discrete dynamical modeling course, which includes the study of linear and nonlinear difference equations; systems of equations; analytic, numeric, and graphic solution methods; conjecturing; analysis of long term behavior; proportionality modeling; and applied problem solving, can accomplish these goals while establishing the mathematical foundation of complexity theory and information science. Throughout such a course, major mathematical themes can be studied. These themes include using functions; investigating the limit process; examining change; examining accumulation; performing approximation; visualizing relations using graphs; developing and analyzing models; generalizing; and performing solution methods.

How do you do this? Undergraduate students need to transition from high school thinking to the higher level of college learning. An introductory discrete dynamics course is perfect for performing that transition. It can be integrated into a core program containing additional topics in other subjects. In order to provide this opportunity by changing current curricula, the content of existing courses may need to be modified. Efficient computation and covering fewer analytic techniques can help find room for dynamical modeling topics. Application problems and larger lively interdisciplinary projects require students to model and analyze the dynamic behavior of monetary accounts, voting trends, market shares, populations, biological systems with predators and prey, chemical reactions, and heat flow. A first-year, core discrete dynamical modeling course is accessible, valuable, and beneficial for all students, even those who will not continue their study of mathematics. [4,34,38]

Inquiry and Modeling

The growth of technology, basic science, and mathematics has been intimately intertwined throughout the history of human civilization. The relationships involved have traditionally offered mutual support and stimulation. But as each of these three broad areas of activity has grown in diversity and complexity, unintended barriers to communication and cross-fertilization have occasionally arisen.

— Frank H. Stillinger (in *Preserving Strength While Meeting Challenges*, 1997)

Because student growth and maturity are important in the context of a core program, attention should be given to the development of student abilities and attitudes in areas such as thinking, reasoning, problem solving, computing, and communicating [2]. A course in inquiry and modeling is ideally suited to develop these important abilities and attitudes in our students. Some of the natural opportunities for development of these attributes through this course are as follows:

- **Inquiring and Reasoning.** Conjecturing solutions and verifying their accuracy are natural processes in a first course in dynamical systems. Often multiple representations (analytic, graphic, numeric, word) of structures are possible. Performing analysis by generalizing concepts from specific examples is also an important part of this course.
- **Modeling and Problem Solving.** Modeling is ideally suited for introducing the fundamental concepts of scientific problem solving [2]. Performing the mathematical modeling process of making assumptions, building models, solving models, and verifying the conclusions, contributes to understanding mathematics and its application. Modeling is used to predict or explain changing behavior (for example the discrete dynamical modeling topics described previously), such as proportionality or linear growth or decay. Further refinements produce nonlinear equations or systems of equations. Students can solve applied problems using their skills in modeling, computing, and reasoning. These applications provide motivation for students by showing them the relevance of mathematics in their future lives.
- **Computing.** Understanding the roles, capabilities, and limitations of technology is critical to student success. In this course, students can use computer software (especially computer algebra systems and spreadsheets) or calculators as tools for iterating, computing, exploring, visualizing, graphing, solving, simplifying, and integrating various means of problem solving. Recent developments in the calculators and computing software make analysis of systems of equations easily accessible to undergraduate students. Computers and calculators are natural tools to help students in most stages of modeling.
- **Communicating.** A modeling and inquiry course provides many opportunities for students to grow in their communications skills—expressing ideas clearly and effectively using proper mathematical notation.

Next Step in Core Mathematics?

Every person who has a liberal education ought to be at some level technologically literate, and it's our responsibility to provide the opportunity for that to happen.
— William Wulf, President of the National Academy of Engineering (November 1999)

What should be the core mathematics program as we enter the twenty-first century and the information age? What should we teach and how should we teach it?

What we teach. Mathematics is becoming universal, and many subjects involve the use of mathematical modeling and inquiry. This *mathematization* of society makes core mathematics an important part of an undergraduate education. It is imperative that our nation's colleges design and implement innovative curricula that contain the content to integrate important topics, along with developing skills in using technology, solving problems, reasoning logically, and understanding the sciences and other disciplines. Also needed are interdisciplinary experiences that give students the opportunity to connect their mathematics, especially modeling, to real problems involving aspects of many disciplines. It is more important to teach processes (problem solving and thinking — modeling and inquiry) than information or techniques (available through technology). The core curriculum needs to be tied together with student growth threads (e.g., reasoning, communicating) — these threads bind together the content among all the required courses as well as form the basis for the development of important student attitudes and skills. This core foundation further affords opportunities for undergraduates to progress in their development as life-long learners who are able to formulate questions, research answers, reach logical conclusions, communicate, work on collaborative teams, make informed decisions, and study quantitative-based disciplines, such as business, science, engineering, and economics. The undergraduate core program must combine the art, language, science, and problem-solving aspects of modern mathematics. This foundation is critical in the development of the future citizen for the highly technical world of the twenty-first century.

How we teach. Core-course concepts should be constantly interconnected and applied to representative problems from business; professional subjects; computing; physical, social, behavioral, earth, and life sciences; and engineering. Interdisciplinary application problems solved using teamwork are valuable to develop student experience in the use of technology, problem solving strategies, mathematical modeling, scientific reasoning, and technical communication skills (written and oral). In active classrooms, students develop a curious and experimental disposition; perform critical and creative thinking; and effectively communicate their ideas and results. These problem-solving activities must be performed in disciplinary, interdisciplinary, multidisciplinary, individual, and team settings for progress to be made in the myriad of requirements confronted by our college graduates during their careers. The technology will change the textbook. Hyperlinks and embedded multi-media modules will change the way students read, learn, study, and write.

Conclusion

All the pictures that science now draws of nature ... are mathematical pictures.

— Sir James Hopwood Jeans (1930)

Since our educational programs and courses should be designed for the future needs of students, we propose that course(s) in inquiry and modeling (i.e., discrete dynamical modeling) enter the core as required mathematics courses for most undergraduate students. These types of courses would prepare students for success in the information age through the understanding of the basic underpinnings of the new sciences and through development of the human role in the information age. These courses provide opportunities for students to mature and grow in solving problems and modeling behavior. Students graduating from programs based on inquiry and modeling will possess the thinking, reasoning, modeling, computation, and language of the new sciences of the twenty-first century and the modeling and problem-solving skills to lead society.

Integrated Core Program

If science is viewed as an industrial establishment, then mathematics is an associated power plant which feeds a certain kind of indispensable energy into the establishment.

— Salomon Bochner

Introduction

Departments of Mathematics across the nation are challenged to modify existing core programs or to develop new ones to meet the changing (mathematical) needs of partner disciplines, the changing components of student growth, and the changing societal needs arising from globalization and our information society. There is an increased emphasis placed on interdisciplinary cooperation – coordination of syllabi, selection of curriculum topics, interdisciplinary student projects, team teaching, and faculty collaboration. Advances in technology for teaching and learning have lowered the access barriers for many topics and rendered the teaching of numerous algorithmic techniques obsolete. Assessment and validation of programs is shifting from prescribing content to focusing on outcomes. For example, the Accrediting Board in Engineering and Technology (ABET) has replaced its specifications of courses and course hours with outcome measures. The rapidly expanding knowledge base and the advancing technology for communication places a much greater emphasis on inquiry and the means of inquiry.

Core mathematics education is education for all students. The development of thought processes judged fundamental to understanding of basic ideas in mathematics, science, and engineering and how they can be applied is critical for the advancement of our society. This is sometimes called quantitative literacy. Robert Witte, the Senior Program Officer for the Exxon Education Foundation, calls this becoming "science savvy."

> "While we remain interested in having well-prepared scientists and engineers for our business, we generally believe that the higher education community does that well. A much more pressing concern is the "science savvy" citizen issue—persons cannot be responsible citizens in today's world unless they have much more profound and fundamental understanding of math and science. It is not enough or even desirable, perhaps, to know science facts. Knowledge of the processes of science, their powers and limitations, is what is needed. This is a major shift for universities, from the goal of preparing a few competent scientists to the goal of every student being scientifically and mathematically literate."*

The emphasis of the core program is at the conceptual level, where the goal is for students to internalize the unifying framework of mathematical concepts. Concepts are applied to representative problems from

* Correspondence with the editor.

mathematics, science, engineering, and the social sciences. These applications develop student experiences in modeling, show the interdisciplinary role of mathematics, and provide motivation for developing a sound mathematical foundation for future studies. Additionally, these applications introduce interests for potential further study.

Inherently, progressive student development underscores that education is a relatively inefficient process and that student time must be provided for experimentation, discovery, and reflection. Viewed from this perspective, the core mathematics experience is not a terminal process wherein a requisite subset of mathematics knowledge is mastered. Rather, it is a vital step in an educational process that enables the student to acquire more sophisticated knowledge more independently. The student who successfully completes the core mathematics program should possess a curious and experimental disposition, have the scholarship to formulate intelligent questions, seek appropriate references, and independently and interactively research answers.

What is an Integrated Core Program?

An integrated program focuses on developing themes by integrating the treatment of appropriate topics drawn from a collection of courses. The first step in developing an integrated program is to identify one or more unifying themes that will provide cohesiveness and direction to the program. Modeling, change, accumulation, approximation and error bounds, transformations, and relations are examples of possible themes. The next step is to identify desired outcomes with respect to content, student growth, and societal needs. These are called end-states. An example is modeling a multiple source pollution problem with a system of differential equations, solving the model, and communicating the interpreted results in both a written and verbal format. The third step is to select topics and applications related to the theme(s) and to the development of the end-states. The topics may be drawn from several courses and the applications may link the program to ideas and concepts in partner disciplines. The final step is to integrate the treatment of these topics into a unified program. Thus an integrated program combines the treatment of major topics from a collection of courses to focus on the development end-states related to curriculum themes. This is in contrast to a traditional course that is designed to develop a single subject such as calculus or linear algebra. The time span of an integrated program is usually less than the collective time span of the individual courses from which the topics are selected. For example, topics addressing a change theme drawn from differential and integral calculus, linear algebra, and differential equations could be integrated into a two-course program. This would be called a "4-into-2" program (topics drawn from four courses integrated into a two course sequence). In addition to the content topics, an integrated program incorporates components of student growth and societal curriculum demands into the development of the end-states based on the curriculum theme(s). An integrated program can be viewed in three input areas: content, student growth, and societal demands.

- **Content.** Topics are selected to address end-states based on curriculum theme(s). For example, to address the end-state of modeling multiple source pollution problems with a system of differential equations, solving the model, and communicating the interpreted results in both a written and verbal format could involve the following ordering of topics:

  discrete dynamical systems (DDSs) → systems of DDSs
  → matrix algebra → (eigenvalue-eigenvector solutions)
  → difference equations → derivatives
  → differential equations → systems of differential equations.

  If time permitted, considering linear versus nonlinear change and/or deterministic versus stochastic change could be included to enrich this sequence.

  There are often greater depth and less breadth in an integrated program than in a sequence of traditional courses. The content is controlled by the unifying themes, which provide cohesiveness and direction to the program.

- **Student Growth.** Student growth in areas of inquiry and modeling (problem solving) is an important part of any mathematical program. The importance of student growth requires that time is made available for inquiry activities and group work. Furthermore, goals need to be explicitly stated, strategies developed, and assessment methods devised to guide the growth. Student growth in these areas is too important to be left to chance.
- **Societal Demand.** The rapidly changing world environment driven by cultural issues and technological advances requires programs that are flexible and responsive to the needs of business and society. Five examples of these needs are

1. Communication skills (reading, writing, presenting, listening)
2. Experience in effectively working as a team member to solve realistic problems
3. Experience in connecting mathematics to other disciplines (building quantitative literacy and interdisciplinary perspectives)
4. Thinking and acting creatively
5. Developing the capacity and willingness to pursue progressive and continued educational development

Technology has connected the world and is a strong driving force for change throughout society, including academia. Awareness and an appreciation for the role of technology in society must underlie future curriculum development.

Components of an Integrated Core Curriculum

> *Computers can never eliminate the need for problem solving through human ingenuity and intelligence.*
> — Paul Brock

Program Goals. Specific goals need to be stated for each of the content, student growth, and societal components of the program. An example for a two semester core program is:

- **Content goals.** conceptual understanding of rate of change (discrete and continuous) and how rates of change are applied in partner disciplines (this includes solving differential equations). (Partners could be science, engineering, life sciences, etc.)
- **Student growth goals.** skill in inquiry and modeling; develop competent, confident, and creative problem solvers.
- **Societal goals.** any or all of those previously listed under the societal demands (e.g., communications, quantitative literacy, creativity).

Course Guide. A course guide is often necessary for an integrated program. The guide provides unification and direction to the program, tasks that are often associated with a textbook. Because topics are chosen from different courses, there is usually not a single suitable text for the program. Thus the need for a guide to establish connections between topics, student activities, projects, and the content theme(s) in order to safeguard the program from becoming a topics course. The course guide can be a "text guide" by containing lesson objectives, study questions, suggested student activities, group projects, and the syllabus. The textual material for some lessons may be contained in the course guide, while for other lessons a text or other resource material, possibly Web-based, will supplement the course guide. (Some schools require students to purchase one or more texts to be used with the course guide.) The course guide provides flexibility to the program as it can be easily modified from semester to semester.

Student Activities. Student activities designed to address specific program goals are scheduled throughout the course. Among the most important are the inquiry-based activities. (These are sometimes called discovery, developmental, or research activities.) Their importance lies in their support of students learning how to learn without formal instructions. This skill is central to developing life-long learners and is essential for students to generalize experiences to gain conceptual understanding.

Example of an inquiry based student activity:

> Make up three second-order, homogeneous differential equations, one having distinct real eigenvalues, one having repeated eigenvalues, and one having complex eigenvalues. Determine scenarios that would result in each of these models. Plot the direction fields for each of the differential equations and, for the first two examples, superimpose the plots of the eigenvectors. Explain the roles of eigenvalues and eigenvectors in determining long term behavior.

Group Projects. Projects offer good opportunities to address the major aspects of student growth and societal demands. Inquiry, problem solving, working as a team member, and communicating through written reports and class presentations are major components of group projects. In addition, interdisciplinary projects offer opportunities for interdisciplinary cooperation at the faculty level and an expanded spectrum of applications at the student level. Project INTERMATH, directed by the U.S. Military Academy, has developed a large collection of Interdisciplinary Lively Application Projects (ILAPs). ILAPs, developed by interdisciplinary teams of faculty, provide students with experiences in solving interdisciplinary problems as well as illustrating the relevancy of mathematics in partner disciplines. These projects usually involve 8–10 hours of "team time" and culminate in a written report.

Technology. Learning how to use technology effectively for learning is an important aspect of a core curriculum. Learning to use a calculator or computer as an inquiry tool as well as a visualization and computational tool is an important component in student growth. Each of the three system environments – graphic, numeric, and symbolic—provide insight opportunities that are difficult to find in the other two environments. Technology can lower the access barriers to content and provide for efficiency in the learning and development components of core courses. This is particularly true for qualitative analysis that is made possible through visualization using technology. Graphically approximating the zeros of a function and illustrating convergence of Taylor polynomials are two examples. Another example is the analysis of long term behavior from direction fields and eigenvector plots, as described previously.

Assessment. Kathi Snook, speaking on assessment at an interdisciplinary workshop in November 1999, said

> As we change how and what we teach, we must change how and what we assess.... Assessments can include a combination of modeling, problem solving, writing, producing or analyzing multiple representations, technology, analytic calculations, or symbolic manipulation. They can be in the form of quizzes, exams, essays, projects, problem sets, or journals. Assessment should reflect the methods and approaches the instructor has posed in the classroom. Results from assessment plans that use a variety of problem types and problem presentations offer a comprehensive view of students' understanding.

Developing conceptual understanding is a highly nonlinear process compared to the more linear process of developing procedural knowledge. Thus assessment practices need to reflect this nonlinearity.

Inquiry and Modeling

Introduction

One approach to developing a mathematics program responsive to the changing mathematical needs of partner disciplines and the work place as well as advances in learning theory is to emphasize the development of inquiry and modeling skills. Because inquiry and modeling can be addressed at all levels, the development of these skills can serve as the goal providing direction for non-major programs as well as major programs. These skills, performed in the context of practical applications, can effectively build quantitative literacy for all students [52].

Calculus programs have provided the direction for many undergraduate core mathematics curricula for at least the past fifty years. However, the changing demands on the core program both from within academia and from society call for a change in the core, one that will better encompass student growth, expand connections with other disciplines, and provide greater depth in problem solving.

Prior to 1990, most core curricula focused on continuous change. The curriculum consisted of a collection of courses, usually calculus I, II, III followed by differential equations. In many liberal arts schools, linear algebra was the fourth course instead of differential equations. If a school only had two core courses, they were often calculus I and II. If there were three courses, then they would be calculus I, II, and III. Lecture was the pedagogical style with emphasis placed on accumulating facts, computation, rigor, and theoretical understanding.

In the decade of the 1990s, core curricula underwent significant change in both content focus and pedagogy. The content expanded to include discrete, as well as continuous, change. There were two major approaches to doing this. One was to integrate the treatment of one and several variables, replacing the standard three-semester calculus sequence with a two-semester calculus sequence [43]. This allowed space for courses in linear algebra and/or discrete mathematics in the core program. The other major approach was to form an integrated program by relating topics from several courses to a unifying theme. For example, at the U.S. Military Academy, topics from seven courses: calculus I, II, III, linear algebra, discrete mathematics, differential equations, and probability and statistics were molded into an integrated four-course program (7 into 4). Topics were selected to address a change theme: discrete and continuous, linear and nonlinear, and deterministic and stochastic. The idea of integrating topics from several courses represented a fundamental change in curriculum development from the previous practice of aligning existing courses. [15]

Pedagogical changes brought about by the Calculus Reform Movement are clearly evident in today's core programs. Emphasis shifted from teacher-centered lectures to student-centered constructivism, communication, realistic applications, group work, active classroom, and learning how to learn. The changes from lecture-directed classrooms to student-directed classrooms can pose time conflicts. These conflicts are accentuated by including group projects such as the ILAPs that are used to link the mathematics department to partner departments and to motivate students (see [3] and the Appendices for this volume). Although eliminating topics is an answer to the time conflict, doing so is extremely difficult. This is particularly true in the calculus portion of the curriculum where "sacred cows" abound. Integrating topics within the core program provides an alternative answer to the time conflict issue.

Currently core programs are undergoing a second significant content change to address the needs of our evolving information society. Inclusion of discrete structures is one response to the growing fields of information technology and data analysis. Some existing topics will need to be removed in order to make room for new topics. Most of the elimination will come from calculus related topics. Other content changes will result from adopting an interdisciplinary structure for the curriculum compared to the present "add on" interdisciplinary component represented by projects.

On the pedagogical side, the mid-1990s debate over whether or not to use technology for teaching and learning has given way to questions concerning how to use technology effectively and appropriately and what technologies are most appropriate to use. Other pedagogical changes are being driven by additional questions such as:

a. How to use technology effectively in teaching students who have grown up with technology?
b. How to respond to the ever-increasing number of "core topics"?
c. How to respond to corporate America's demand for competent, confident, and creative problem solvers?
d. How to filter meaningful information from an ever-increasing amount of data and how to extract knowledge from that information?

The need for broader content coverage and the advances in technology for researching information (Web, electronic libraries, on-line texts, email) combine to emphasize the central importance of inquiry in the core curriculum. Further underscoring this emphasis is the fact that advances in technology and the spirit of entrepreneurship that have driven the globalization of the economy have also raised demands for students to be better able to learn on their own. These demands, in addition to emphasizing inquiry, strongly suggest that modeling be made a central focus of the core program.

The on-going explosion of information and the ever-increasing demand for skilled problem solvers calls for a core program focused on developing inquiry and modeling skills. The implementation of such a program may diminish the role of calculus and increase the roles of data analysis, discrete mathematics, interdisciplinary cooperation, and student growth.

Inquiry and Modeling

An appropriate pedagogical response to an ever-increasing assortment of "core topics" and society's demand for problem solvers is to emphasize inquiry and modeling. A basic assumption behind this response is that skills in inquiry and modeling better develop students in the following aspects than have the traditional courses in the past. Students should

a. learn how to learn and become life long learners,
b. generalize processes and results,
c. draw connections between topics and disciplines, and
d. gain an appreciation of the role of mathematics in society.

A consequence of this assumption is that studying a few concepts in depth, by studying related topics in several courses, better prepares the student for tomorrow's world than does focusing on a few traditional courses.

Inquiry has been a critical component in learning since learning began. How do we formulate a question? How do we move from a question toward a solution? How do we generalize a result? How do we develop conceptual understanding? Although there may be algorithmic like processes for guiding an inquiry into a particular topic, there is not a body of material called "inquiry" that can be mastered. Instead there are a few general principles (e.g., understand the statement of the problem) and many specialized activities (e.g., analyzing alternatives, sorting out questions like *what if? so what?* and *now what?*). Because inquiry is not a content topic and because inquiry is essential to the art of learning, core programs must make inquiry explicit in the curriculum. Five suggestions for doing this are:

1. Assign new material for students to prepare (including a few exercises) before it is discussed in class.
2. Emphasize the practice of analyzing alternatives. Exercises where students ask *what if*, *so what*, or *now what* and seek out the alternatives can develop confidence and lead the students to distinguish between

problems and exercises. Extending the *questioning* technique to problem situations leads the student to identify and then address the underlying "why" type questions, an activity that is central to gaining conceptual understanding.

3. Raise students' awareness level of inquiry by explicitly noting, discussing, and illustrating methods of inquiry as part of classroom lessons.
4. Conduct a biweekly inquiry driven, two-part problem-solving lab. The first part is done in class and consists of students working in teams on a discovery type problem. The second part is a follow-up homework assignment to research some aspect of the problem. For example, write an essay on a historical component of the problem or refine the group analysis to find a better solution.
5. Begin class by asking students to state questions they developed based on their homework. Although the majority of responses probably will fall into the mechanical category associated with a particular exercise, there will be some (possibly supplied by the instructor) that fall into the curiosity, dream, or generalization categories. By discussing and emphasizing these types of questions, the instructor can help move students into an inquiry environment.

Modeling is composed of three stages as illustrated by the legs of the triangle in the modeling diagram. The Model Construction leg involves identifying the situation to be modeled and question(s) to be answered, identifying variables and relations between them, and making assumptions. Because real world situations are too complex to allow the modeler to account for every facet of the situation, simplifying assumptions arc made to obtain a "first" model. For example, in a savings account model months are assumed to have equal length. Or, in projectile models air resistance is usually ignored. Subsequent refinement of a model is developed, in part, by altering the assumptions. Data collection and curve fitting are often part of model construction. Graphs, functions, differential equations, systems of equations are examples of forms of mathematical models. Most of the exercises in a calculus text can be viewed as models (often stated in isolation of any meaningful context).

The major emphasis in most first and second year mathematics courses is focused on the Solution Techniques leg of the modeling diagram. Because of the algorithmic nature of solution techniques, modern technology can accomplish many of the needs of this leg. As a result, technology has lowered the access barriers of several topics and provided opportunities for reforming curricula.

The Interpretation leg involves determining both the suitability and reasonableness of a mathematical solution. Because assumptions are made in constructing a model, the resulting mathematical solution is often an approximate solution to the original problem. Thus suitability questions exist concerning the level of accuracy of the approximate solution. Also a correct solution to a mathematical model is not necessarily a reasonable solution to the problem being modeled. Consider the problem of determining the time required for a ball to hit the ground when dropped from the top of a building. One approach is to model the position of the ball as a function of time and then solve for the zeros of the function. The function is

Modeling

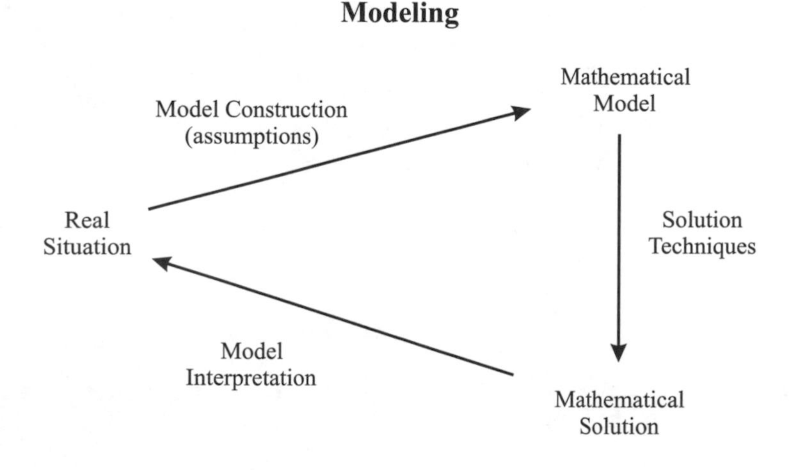

quadratic with one positive zero and one negative zero. The positive zero is the time when the object hits the ground. The negative zero, although a correct mathematical solution, is not a reasonable solution to the problem.

Example. Determine the times over a three-day period when the oxygen level in Bog Stream is 25 ppm based on the following four samples taken the first day: 12 ppm at 1:00 AM; 20 ppm at 7 AM; 36 ppm at 1:00 PM; 22 ppm at 8:00 PM.

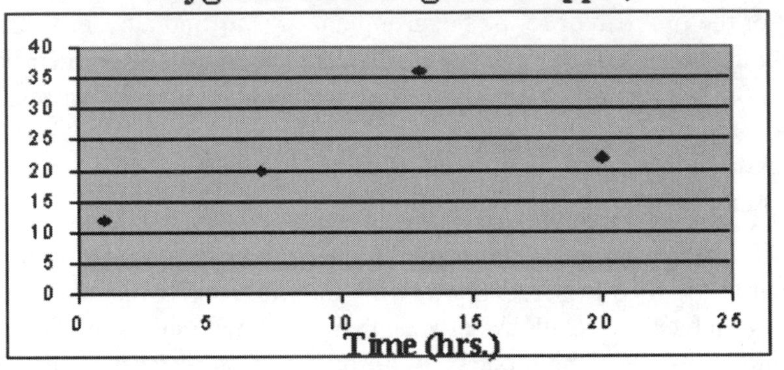

There are several curves that could be used to fit this data. Because four points uniquely determine a cubic equation, we might begin with a cubic regression program to obtain the cubic model. Setting this cubic expression equal to 25 gives the mathematical solutions 8:74 AM and 7:65 PM (on the first day only).

The accuracy of the solutions is acceptable for the first day, but the model is not suitable as it shows negative values of oxygen during the second and third days. This interpretation of the mathematical solution shows that a different model needs to be considered. Is a periodic model a reasonable choice to make? The answer depends on understanding what causes oxygen levels to fluctuate. Thus an inquiry activity into the factors effecting oxygen levels is needed before continuing the modeling process.

The Construction and Interpretation legs of the modeling process often require greater conceptual understanding and skill in inquiry than does the Solution Techniques leg. Traditionally problem solving only applied to the Solution Techniques leg of the modeling process. In focusing the core program on inquiry and modeling, we are expanding the concept of problem solving to include all three legs of the modeling process.

Framework for a Two-Semester Integrated Core Program

The following block-outline is for a two semester (100 lesson) integrated core program based on a change theme — discrete and continuous. The desired end-state is the development of competent, confident, and creative problem solvers; able to model discrete and continuous change situations. Topics are drawn from graph theory, discrete mathematics, linear algebra, calculus I & II, and differential equations. Topics are introduced through realistic situations and developed using inquiry and modeling. The acronym *PSL* refers to a small group problem-solving lab (conducted in the classroom). The program depends on extensive use of technology, in particular visualization. The program also assumes that students are expected to encounter and to develop new material in preparation for class before the material is discussed in class.

Iteration is a basic process in the program. Iterations are performed to conjecture long-term behavior, discover patterns, and provide insight. For example, iteration is the process that leads to the development of symbolic solutions of discrete dynamical systems (DDSs). In the calculus portion, iteration is generalized to form a sequence of successive approximations. These form the basis for developing the definitions of derivative, integral, improper integral, and Taylor polynomials.

The first block introduces graph theoretic modeling in discrete, static situations. The second block extends the modeling to discrete change. Discrete dynamical models are developed in block 4. Modeling of continuous rates of change begins in block 7. The analysis of long-term behavior in blocks 2, 4, and 5 provides the basis for limits of sequences used in the sequential approach to calculus starting in block 7. Differential equations are first introduced as a means of *undoing* the differentiation operation (i.e., antiderivative). Euler's Method is developed in connection with the derivative and antiderivative. Integration is introduced as accumulation, distinct from antidifferentiation.

The description of each block includes a statement of purpose, a topic flow diagram, and some specific content suggestions. The purpose of thc annotated block-outline is to provide a flavor and demonstrate existence of an integrated program, not to specify a particular syllabus.

Block 1 (Lessons #1–8): Introduction to graph theoretic models of static situations.

Purpose: Introduce students to modeling using graph theoretic models and introduce matrix notation to represent graphs and networks.

Topic Flow: Graphs → Euler Circuits → Hamilton Circuits
→ Spanning Trees → Network Paths → State Diagrams.

Suggestions: Begin the block with a PSL on the Konigsberg Bridge problem. Subsequent lessons generalize the graph model of the Konigsberg Bridge problem, illustrating how a model can provide information and insight to the particular problem as well as to related problems. Introduce adjacency matrices and matrix multiplication to represent paths of length n.

Block 2 (Lessons #9–18): Long-term behavior for probabilistic models of dynamic situations.

Purpose: Introduce probabilistic models (Markov Chains) to describe change between different states. Introduce the question of long-term behavior and then address it by iterating a transition matrix. Graphically discover eigenvalues and eigenvectors. Find the eigenvalue-eigenvector solution.

Topic Flow: Adjacency Matrix → Transition Matrix (via probability) → Matrix Properties → Iteration → Eigenvalue & Eigenvector Solution.

Suggestions: Introduce concept of probability in a simple and intuitive way. Generalize a graph or network by labeling the edges with probabilities, producing a state diagram and then generalize an adjacency matrix to a transition matrix (e.g., use a rat's maze). Short PSL on the long-term result of a rat running through a maze. Or, PSL: Model Margaret's choice of ice cream—if Margaret's last ice cream was blueberry, she will order blueberry again 70% of the time and some other flavor 30% of the time. If her last ice cream was not blueberry, she will order blueberry 40% of the time and some other flavor 60% of the time. Model the situation with a state diagram and then a transition matrix. Select a starting state and then iterate it six times with the transition matrix. Conjecture long term percentage of time that Margaret orders blueberry. Check the conjecture by using technology to iterate 50 or 100 times.

In subsequent lessons, view a matrix as a function mapping a vector into a vector. Consider this mapping graphically and symbolically. Graphically discover eigenvectors and eigenvalues by iterating a transition matrix on a vector. Decompose a 2-dimensional vector into two vectors by *undoing* the graphical representation of adding two vectors (Parallelogram Law). All the pieces now exist for representing the kth iterate of a transition matrix on an initial vector as a linear combination of the kth powers of eigenvalues times their corresponding eigenvectors. This is a great accomplishment and time should be taken to relish it by modeling several situations that lead to transition matrices.

Note: Present vectors as a generalization of real numbers to higher dimensions—an ordered n-tuple. Do not attempt to introduce vector spaces.

Block 3 (Lessons #19, 20): Review and Test.

Block 4 (Lessons #21–28): First order discrete dynamical systems (DDS)

Purpose: Model dynamic situations that change at discrete time intervals. Understand and apply the conjecture-validate process. Develop a solution process based on iteration. Analyze long-term behavior of a DDS. Understand how the paradigm (New Situation) = (Old Situation) + (Change) is used as the basis for modeling DDS situations.

Topic Flow: Homogeneous DDS → Nonhomogeneous DDS, both constant & exponential

Suggestions: PSL: Develop a DDS model for an investment program that consists of investing an initial amount with a fixed, monthly interest rate. Use technology to iterate the DDS in order to determine the balance after a specified number of months. Iterate by hand to recognize a pattern and then conjecture the balance after k months based on the observed pattern. Verify the conjecture by showing that it satisfies the DDS. Emphasize five things:

1. Defining variables and stating assumptions are important in constructing the DDS model.
2. A numerical solution (for the balance after k months) can be obtained by iteration.
3. Arithmetic operations (addition, multiplication) performed when iterating can disguise patterns and should not be performed.
4. The process for verifying a conjecture is important and must be understood.
5. This process may help understand the long-term behavior.

Investigate alternatives of using different time periods and including different fixed monthly deposits. Expand the scope of the model to include situations involving credit card debt, car payments, mortgages, annuities, drug dosage, etc.

Note: Iteration of a DDS generates a finite geometric series. Recognizing this series is a key element in recognizing the DDS pattern.

Reserve time at the end of this block to reflect on the meaning and applicability of a DDS. In particular, stress the structure of a DDS

(New Situation) = (Old Situation) + (Change)

ILAP (in coordination with Department of Economics): "Car Financing" by Richard West, Jeffery Durnford, Scott Torgerson, and Michael Roane, COMAP, 1995.* This ILAP involves analyzing several different options for financing a new car.

Block 5 (Lessons #29–38): Model systems of first order DDSs

Purpose: Understand how systems of first order DDSs arise. Understand how an eigenvalue-eigenvector solution is found. Understand the role of eigenvalues in analyzing long term behavior.

Topic Flow: System of DDSs → Eigenvalue/Eigenvector Solutions → Higher Order DDSs

Suggestions: PSL: Model a transition matrix situation (Block 2) and then model a predator/prey situation. Emphasize the five items from the previous block.

Model realistic situations that lead to a homogeneous system of first order DDSs. Interpret the role of the eigenvalues in analyzing long term behavior.

Generalize from the homogeneous to the nonhomogeneous case by recalling the situation with a single DDS.

Model the number of ways of climbing a flight of n stairs if a person covers one or two stair treads with each step. Transform this second order DDS into a system and solve. Repeat with other situations yielding a second or third order DDS. Use understanding of the system's matrix characteristic equation/eigenvalues to develop the characteric equation for a higher order DDSs.

ILAP (with Department of Chemistry) "SMOG in Los Angeles Basin" (see [3]).

Block 6 (Lessons #39, 40): Review and Test

Block 7 (Lessons #41–48): Model continuous rate of change.

Purpose: Understand continuous rate of change. Develop the derivative concept in multiple ways:

Limit of sequence of successive approximations, i.e., average rates of change as the interval length approachs zero

Geometric: secant lines → tangent lines

Numerical

Time period approachs zero in a DDS

Topic Flow: Approx. Continuous Change → Derivative → Properties → Antiderivative

Suggestions: Student groups collect data by conducting distance/time rate experiments (e.g., rolling a tennis ball down the hall, driving a car). Analyze the plot of the data.

Note similarity of investigating long term behavior of a DDS and the convergence of a sequence of successive approximations. (The work with DDSs provides a strong basis for the sequence approach to calculus.)

Emphasize:

Qualitative analysis of basic functions (shapes, increasing, concavity, zeros, etc.)

Linear Approximation

---

* Available at website `http://www.projectintermath.org/products/listing.`

Differentiation rules
Physical interpretations and applications of the derivative concept (optimization).

Block 8 (Lessons #49, 50): Review and Final Examination

Block 9 (Lessons #51–56): Review of Differentiation, Optimization, and Anti-differentiation

Purpose: Use the review of differentiation to introduce differential equations as anti-derivative problems. Develop Euler's Method and solve easy differential equation problems (e.g., separation of variables).

Topic Flow: Review differentiation antiderivatives Euler's Method

Suggestions: Rely on graphics to lead and inform the analysis. Optimization—emphasize the modeling aspect, rely on qualitative analysis to lead to the analytic techniques for optimization. PSL: Introduce slope fields and solution curves. Students explore differentiating (simple) function to get a differential equation, sketch the corresponding slope field, sketch solution curves and compare them to plot of the function. Given the plot of a function, sketch the corresponding slope field.

ILAP (with Department of Physical Education) "Getting Fit with Mathematics" (see [3]).

Block 10 (Lessons #57–70): Accumulation and Numerical Integration

Purpose: Develop integration concept as accumulation using the sequence of successive approximations approach. Understand the Fundamental Theorem of Calculus.

Topic Flow: Approximations Numerical Integration Integration Fundamental Theorem

Suggestions: Students develop reasonable approximation to the area of a region (e.g., pond, irregular shaped field, irregular shaped parking lot), make two successive improvements in their approximation, and develop a procedure for systemically improving their approximation. Note similarity of investigating long term behavior of a DDS and the convergence of a sequence of successive approximations to the area of the region.

Apply the accumulation concept in several different settings in addition to area. Use technology for computing integrals.

Block 11 (Lessons #71, 72): Review and Test

Block 12 (Lessons # 73–98): Differential Equations

Topic Flow: Direction Fields → Euler's Method → Separation of Variables → Systems (Predator-Prey) → Qualitative Solutions → Forced Harmonic Oscillators → Qualitative Solutions

Suggestions: Review slope fields, solution curves, and Euler's method. Note the parallels between the study of DDSs and differential equations. Use the DDS experience to inform the study of differential equations.

Model situations via first-order systems, express model in matrix form, solve using eigenvalues and eigenvectors (i.e., mimic the DDS development). Focus on qualitative solutions. Conclude section with the study of harmonic oscillators.

ILAP "Flying with Differential Equations" (see [3]).

Block 13 (Lessons #99, 100): Review and Final Examination

Creating an Environment for Change

The chief forms of beauty are order and symmetry and precision that the mathematical sciences demonstrate in a special degree.
— Aristotle

Technological advances continue to challenge us in terms of how we access information, how we learn, and how we compute. Technology has lowered barriers to content and thus has lessened the hierarchical structure of a mathematics curriculum; while computer algebra systems (CAS) have freed us from the need to concentrate our teaching on algorithmic manipulation. These changes are initiating a reexamination of curriculum goals and course content. Today, our information society places a much greater emphasis on modeling and inquiry than it did ten years ago. The development of technology for teaching and learning initiated a revolution in our curricula ten years ago and is today forcing us to redefine our programs again. During the 1990s we reacted to these issues and played catch up in adapting curricula to the capabilities of graphing calculators and CAS on computers. Now, before we have finished adapting to these changes, we are faced with another technological advancement with respect to teaching and learning, CAS running on graphing calculators. How long will it be before the next major technological breakthrough occurs? How can we adapt or develop curricula to meet the needs of globalization, information age, information technology, and so on?

How can departments and programs prepare for change? How can the faculty be encouraged to adopt new curricula? The level of interest and the quantity of research being conducted into how people learn has reached a new high. How will results from this research find its way into classrooms? How can we become proactive rather than reactive in developing curricula? In short, how can departments create an environment for change? The following offer suggestions for doing this.

Experimental Courses

One way to develop an environment for change is to promote experimental courses in core programs. Faculty teaching these courses should be encouraged to experiment with the curriculum while addressing the departmental goals for the course. Institutionalizing experimental courses provides a laboratory for change, for developing new courses, as well as for adapting current courses to meet changing situations. The existence of experimental courses in the curriculum is a message that the department expects the curriculum to be evolving and is actively seeking improvement. Collins and Porras in their book *Built to Last* cite the willingness to experiment and change as characteristics of visionary companies [11]. These need to be characteristics of our academic departments as well.

Curriculum Retreats

Retreats provide opportunities for faculty to focus on curriculum issues. This experience is further enriched when done in conjunction with teams of faculty from other schools. Although each team works on its own program, the synergistic effects of group discussions enhance each program. In 1994, West Point held such a Curriculum Retreat for seven schools [3]. Five of those schools are now disseminating major curriculum projects that were initiated at the retreat and implemented over the past few years. In 1998, approximately 60% of the faculty of Carroll College (Montana) attended an off-campus, weekend retreat focused on ILAPs. As a result of that retreat, faculty are now creating and implementing ILAPs across the curriculum.

Curriculum Consortiums

Consortiums provide opportunities for faculty from different institutions with different restrictions to come together to work on a common task. Therefore, it provides a broader source of experience, ideas, and resources to its members and their task. Furthermore, it may provide the "critical mass" necessary to produce change and gain the resources for the process. Examples of successful consortium include the differential equation, the linear algebra, and the Historically Black Colleges and Universities College Algebra Reform consortiums.

Interdisciplinary Cooperation

Changing disciplinary environments benefits from good information flows between partner disciplines. A liaison program offers an avenue for enhancing communication between departments. This may include, for example, coordinating course syllabi, informing colleagues how and when mathematics is used in the partner discipline, and determining the mathematical needs of the partner discipline. Creating ILAPs with faculty in partner disciplines provides an activity to link departments and build interdisciplinary cooperation.

Time and Value

Changing programs and developing new courses, particularly modeling courses takes a great deal of time. Institutions should provide faculty with time and support to attend conferences and to develop programs and materials. Institutions should also clearly demonstrate the value of managing change by rewarding those who are involved. Facilitating change needs to be included in the academic reward structure.

Conclusion

Changes in society, business, and in academics oblige departments to establish environments for change. Helping students anticipate and respond effectively to the uncertainties of a changing technological, social, political, and economic world requires departments to become proactive agents for curricula and pedagogical change.

References

1. Arney, D. C., et al. 1995. "Core Mathematics at the United States Military Academy: Leading into the 21st Century," *PRIMUS*. 5(4), 343–367.
2. Arney, D. C. 1996. "Modeling in the First Two Years," *Proceedings of the Second Modeling Symposium*, 29–36.
3. Arney, D. C. (Editor) 1997. *Interdisciplinary Lively Application Projects* (ILAPs), Washington, DC: Mathematical Association of America.
4. Arney, D.C., Giordano, F., Robertson, J.S. 1999. *Discrete Dynamical Systems: Mathematics, Methods, and Models* (Preliminary Edition), New York: McGraw-Hill.
5. Board on Army Science and Technology 1992. *Star 21: Strategic Technologies for the Army of the Twenty-First Century*, Washington, DC: National Academy Press.
6. Borrelli, Robert, Keller, Robert, and Moody, Michael. 1998. "A Preliminary Plan for Curriculum Change at Harvey Mudd College: *n*-into-four". *Confronting the Core Curriculum: Considering Change in the Undergraduate Mathematics Major*. MAA Notes Number 45, Washington: Mathematical Association of America.
7. Breivik, Patricia Senn, 1998. *Student Learning in the Information Age*, Phoenix: The Oryx Press (American Council on Education).
8. Case, Robert, 2000. "Panel on Cultural Change," MAA Meeting, Washington.
9. Cipra, Barry 1993-1999. *What is Happening in the Mathematical Sciences* (Vols 1–4), Providence: American Mathematical Society.
10. Cohen, P. 1982. *A Calculating People*, Chicago: University of Chicago Press.
11. Collins, J.C. and Porras, J.I., 1994. *Built To Last*, New York: Harper Collins Publishers.
12. COMAP. 1997. *Principles and Practice of Mathematics*. New York: Springer-Verlag.
13. Cooney, Thomas and Hirsch, Christian (Editors), 1990. *Teaching and Learning Mathematics in the 1990s*, Reston, VA: National Council of Teachers of Mathematics.
14. Devlin, Keith. 1997. *Goodbye, Descartes: The End of Logic and the Search for a New Cosmology of the Mind*. New York: John Wiley.
15. Dossey, John (Editor). 1998. *Confronting the Core Curriculum: Considering Change in the Undergraduate Mathematics Major*, MAA Notes Number 45, Washington: Mathematical Association of America.
16. Friedman, T.L. 1999. *The Lexus and The Olive Tree*, New York: Farrar, Straus and Giroux.
17. Kemeny, J.G, Snell, J.L., Thompson, G.L. 1957. *Introduction to Finite Mathematics*, Englewood Cliffs, NJ: Prentice-Hall.

18. Haver, William E. (Editor), 1998. *Calculus: Catalyzing a National Community for Reform: Awards 1987–1995*. Washington, DC: Mathematical Association of America.

19. Hoskin, Keith. 1994. "Textbooks and the Mathematisation of American Reality: The Role of Charles Davies and the US Military Academy at West Point," *Paradigm,* Vol. 13, 11–41.

20. Humphries, George (Editor). 1997. *Future Operations/Future Warfare*, special subject of the *Military Review*. March-April.

21. Johnson, George. 1997. "Researchers on Complexity Ponder What It's All About," *NY Times*, Science Times Section, May 6, C1 and C7.

22. Leinbach, L. Carl et al (Editors) 1991. *The Laboratory Approach to Teaching Calculus*, MAA Notes Number 20, Washington: Mathematical Association of America.

23. Leiss, William. 1990. *Under Technology's Thumb*, Montreal: McGill-Queens University Press.

24. Lukasiewicz, Julius. 1994. *The Ignorance Explosion*, Ottawa: Carleton University Press.

25. Mathematical Sciences Education Board 1989. *Everybody Counts: A Report to the Nation on the Future of Mathematics Education*, Washington: National Academy Press.

26. Mathematical Sciences Education Board 1991. *Moving Beyond Myths: Revitalizing Undergraduate Mathematics*, Washington: National Academy Press.

27. Masuda, Yoneji. 1990. *Managing in the Information Society*, Cambridge, MA: Basil Blackwell.

28. Meyer, Walter (Editor) 1992. *A New Start for College Mathematics*, Lexington,MA: COMAP.

29. Morrison, James L. 1986. *The Best School in the World: West Point, the Pre-Civil War Years, 1833-1866*. Kent, OH: Kent State University Press.

30. National Research Council 1997. *Preserving Strength While Meeting Challenges: Summary Report of a Workshop on Actions for the Mathematical Sciences*, Washington: National Academy Press.

31. National Science Foundation 1996. *Shaping the Future: New Expectations for Undergraduate Education in Science, Mathematics, Engineering, and Technology*.

32. Pólya, G. 1946. *How to Solve It: A New Aspect of Mathematical Method*, Princeton: Princeton University Press.

33. Ralston, A. and Young, G. 1983. *The Future of College Mathematics*, New York: Springer-Verlag.

34. Ralston, A. (Editor). 1989. *Discrete Mathematics in the First Two Years,* MAA Notes Number 15, Washington: Mathematical Association of America.

35. Reigeluth, Charles M. and Garfinkle, Robert J. (Editors) 1994. *Systemic Change in Education*, Engelwood Cliffs: Educational Technology Publications.

36. Reynolds, Barbara E. et al (Editors), 1995. *A Practical Guide to Cooperative Learning in Collegiate Mathematics*, MAA Notes Number 37. Washington: Mathematical Association of America.

37. Roberts, Wayne (Editor) 1996. *Calculus: The Dynamics of Change*, MAA Notes Number 39, Washington, DC: Mathematical Association of America.

38. Sandefur, James. 1993. *Discrete Dynamical Modeling*, New York: Oxford Press.

39. Scharf, J. and Vanisko, M. 1998. "Carroll College Mathematics Curriculum Reform Project," *Confronting the Core Curriculum: Considering Change in the Undergraduate Mathematics Major*, MAA Notes Number 45, Washington: Mathematical Association of America.

40. Schoenfeld, Alan (Editor) 1990. *A Source Book for College Mathematics Teaching*, MAA Report, Washington: Mathematical Association of America.

41. Schoenfeld, Alan (Editor) 1997. *Student Assessment in Calculus*, MAA Notes Number 43, Washington: Mathematical Association of America.

42. Siegfried, John J.; Getz, Malcolm; and Anderson, Kathryn H. 1995. "The Snail's Pace of Innovation in Higher Education," *The Chronicle of Higher Education*, May 19, A56.
43. Small, D.B. and Hosack, J.M. 1980. *Calculus, An Integrated Approach*, New York: McGraw-Hill Publishing Co.
44. Small, D. B. 2000. "Core Mathematics for Engineers, Mathematicians, and Scientists," *IJEE, Vol.* 15, No.6.
45. Smith, David; Porter, Gerald; Leinbach, Carl; and Wenger, Ronald 1988. *Computers and Mathematics: The Use of Computers in Undergraduate Instruction*, MAA Notes Number 9, Washington: Mathematical Association of America.
46. Solow, Anita (Editor), 1994. *Preparing for a New Calculus*, MAA Notes Number 36, Washington: Mathematical Association of America.
47. Sorbi, Andrea (Editor).1997. *Complexity, Logic, and Recursion Theory*, Lecture Notes in Pure and Applied Mathematics, Volume 187, New York: Marcel Dekker.
48. Steen, Lynn Arthur (Editor) 1989. *Calculus for a New Century: A Pump not a Filter*, MAA Notes Number 13, Washington: Mathematical Association of America.
49. Steen, Lynn Arthur (Editor) 1989. *Reshaping College Mathematics: A Project of the Committee on the Undergraduate Program in Mathematics*, MAA Notes Number 13, Washington, DC: Mathematical Association of America.
50. Steen, Lynn Arthur (Editor) 1990. *On the Shoulders of Giants: New Approaches to Numeracy*, Washington, DC: National Academy Press.
51. Steen, Lynn Arthur (Editor) 1992. *Heeding the Call for Change: Suggestions for Curricular Action*, MAA Notes Number 22, Washington, DC: Mathematical Association of America.
52. Steen, Lynn Arthur (Editor) 2001. *Mathematics and Democracy: The Case for Quantitative Literacy*, Washington: National Council on Education and the Disciplines.
53. Svinicki, Marilla D. (Editor) 1999. *Teaching and Learning on the Edge of the Millennium: Building on What We Have Learned*, San Francisco: Jossey-Bass Publishers.
54. Tucker, Alan C. (Editor) 1995. *Models that Work: Case Studies in Effective Undergraduate Mathematics Programs*, MAA Notes Number 38, Washington, DC: Mathematical Association of America.
55. Tucker, Alan C. and Leitzel, James (Editors) 1995. *Assessing Calculus Reform Efforts: A Report to the Community*, MAA Report, Washington: Mathematical Association of America.
56. Tucker, Thomas W. 1990. *Priming the Calculus Pump: Innovations and Resources*, MAA Notes Number 17, Washington: Mathematical Association of America.
57. Ward, David 2000. "Catching the Waves of Change in American Higher Education," *Educause Review*, January/February, 22–30.
58. Wilson, Robin 2000. "The Remaking of Math," *Chronicle of Higher Education,* January 7, 2000.
59. Young, Robert M. 1992. *Excursions in Calculus: An Interplay of the Continuous and the Discrete*, Dolciani Mathematical Expositions Number 13, Washington: Mathematical Association of America.
60. Zimmermann, Walter and Cunningham, Steve, 1991. *Visualization in Teaching and Learning Mathematics*, MAA Notes Number 19, Washington: Mathematical Association of America.

Part 2

Commentary

Interdisciplinary Culture Perspective

Technology Perspective

Goals and Content Perspective

Instructional Techniques Perspective

Interdisciplinary Culture Perspective

This section edited by Gary Krahn.

The state of our academic environment, in particular the Interdisciplinary Culture is of great concern to educators across disciplines. Progress in understanding, developing, and broadening our students is restricted by the barriers between departments and the lack of communication among faculty. The authors of these papers examine the future of core mathematics from an interdisciplinary viewpoint. They address the issues of eliminating barriers, establishing partnerships, and improving core mathematics programs to serve partner disciplines in the development of core students.

The implementation of interdisciplinary cooperation into curricula is seen as a goal of growing importance, and one that faces numerous roadblocks. Thomas Berger's suggestion to view curriculum through outcome goals places a strong value on societal demands. Several of these demands underscore the need and value of interdisciplinary cooperation. For example, preparing students for a diversity of careers in a rapidly changing business climate involves making explicit connections among disciplines. Gary Krahn views interdisciplinary cooperation as being important in developing reasoning and critical thinking skills. He writes, "the double movements of induction–deduction, expanding–contracting, generalization–specialization, interdisciplinary–disciplinary reflect the processes we want students to assimilate." Brian Winkel notes, "we are moving (ever slowly) toward a culture that accepts interdisciplinary approaches to teaching what has been traditionally disciplinary material," however, there is an array of obstacles in the way. Some samples are: system inertia, turf protection, publish or perish syndromes focused on narrow results, entrenched attitudes, and lack of a reward system.

Mathematics is not well integrated into college or university curricula today. There are local exceptions and there is increasing discussion on the future (interdisciplinary) role of mathematics in the curriculum. Brian Winkel suggests that dialogue is the key to successfully integrating mathematics across disciplines. Reforming mathematics instruction to emphasize the process of reasoning is the basis on which Gary Krahn and Thomas Berger would build interdisciplinary cooperation. Bob Fuller addresses developing interdisciplinary cooperation by posing the question: "What metaphor should we use to most adequately convey the goals and intentions of the mathematics across the curriculum movement?" How faculty, particularly in partner disciplines, describe mathematics has a significant influence on how students view mathematics. Consider, for example, what messages these common metaphors convey: "The rest is just mathematics…", "Mathematics is a tool to…", "Mathematics is a language for…" convey to students (and faculty)?

The future role of the traditional approach to calculus is in question because of its narrow focus and its role in solving tomorrow's problems. Brian Winkel suggests replacing calculus with a modeling course in which rates of change and accumulation are the themes. Gary Krahn agrees with the modeling suggestion and argues against teaching the traditional calculus course because of its emphasis on topics over process.

The Development of a More Interdisciplinary Academic Culture

Brian J. Winkel
United States Military Academy

Abstract. Mathematics education must continue to advance toward being integrated throughout the undergraduate curriculum. The needs of our society point to an educational process that is engulfed in interdisciplinary approaches to teaching and learning. This note reflects briefly upon the why and the how of a transition to a more interdisciplinary academic culture. The barriers among academic departments should exist to hold up the building and not to fragment the learning process. It will take effort, support, and compassion to meld integrated learning communities, however, it will be an exciting journey.

Introduction

It is my position that we are moving (ever slowly) toward a culture that accepts interdisciplinary approaches to teaching what has been traditionally disciplinary material. There are obstacles. Some are high, some are superficial, but nevertheless real, e.g., inertia of a system, fiefs and turfs, distance between offices, tenured dead wood, cynics, publish or perish syndromes focused on narrow results, etc. However, the evidence that the movement is growing is seen in the programs offered and accepted by supporting agencies (National Science Foundation, private foundations, and local faculty governance mechanisms); the number of articles written about interdisciplinary initiatives and sessions at national and regional meetings on interdisciplinary efforts; and the ever present awareness of how engineering and science works in the real world with teams from different disciplines.

It is not easy to move forward for it takes energy and time and it is all too easy to slip back into the comfortable past and the disciplinary shells which most of us have known all our lives. We address some of the issues outlined by the conference organizers to support our contention that it is both valuable and appropriate to move in the direction of an interdisciplinary culture. Indeed, Barbara Olds, a member of the Engineering Practices Introductory Course Sequence (EPICS) program and an English teacher at Colorado School of Mines, some years ago wrote an end piece article in the American Society for Engineering Education (ASEE) journal, *PRISM*, in which she offered her own "modest proposal" to do away with all departments. We do not see that happening any time soon (unfortunately), but we do see more and more contacts being established. Hopefully, these contacts will weave a web of connection and then identity among disciplines. This will make an effective teaching environment, which can then produce a powerful, broadly educated, richly exampled, student body—a student body, emerging with skills and experiences to solve tough interdisciplinary problems that confront our societies.

What is the impact of mathematics reform on the partner disciplines?

In some instances, colleagues outside the department of mathematics have stood by and watched how calculus reform has torn apart departments; alienated collegial friends; produced students who cannot differentiate sin(x) by hand; increased mathematics budgets for computers and thus forced cuts in their own budgets; and caused some students to say, "I can't learn calculus with a computer;" or "What do you mean write an essay—this is math class!"

Yet significant reform has taken place in physics, chemistry, and engineering. Chemistry is supported with an NSF funded initiative, modeled after the calculus reform effort. Physicists initiated their own reform when they discovered that after all their *equationing* the students just did not get it. Students continued to believe in the *Road Runner physics* they were conditioned by on Saturday morning TV. The mathematics reform, which has stressed use of technology, has been very attractive to some engineering and science colleagues. They are proponents of reform, mathematics-based engineering and have an interest in playing with parameters to address *what if* questions. Computer algebra systems support these investigations very well and the non-mathematics faculty are meeting mathematics faculty on a common ground in using this software to permit discovery and analysis of more complex systems.

NSF has heavily supported engineering education and new curricula have emerged. As an example consider the Integrated First-Year Curriculum in Science, Engineering, and Mathematics (IFYCSEM) developed at Rose-Hulman Institute of Technology in the late 1980s. In this program all the science, engineering, and mathematics course content is wrapped up into three 12-credit quarter courses in which a team of eight faculty from science, engineering, and mathematics teach a cohort of some 90 students. This IFYCSEM has served as a model and has been modified by other institutions (e.g., Texas A&M University and University of Alabama) in the Foundation Coalition, one of several multi-million dollar engineering education coalitions sponsored by NSF.

Furthermore, mathematics faculty members are visiting with engineering and science faculty for ideas to enrich their class and applications to present as projects. These are not just visits to see what they want us to do in mathematics, but rather true collegial exchanges. An example of this is our work with Ed Mottel, chemistry professor at Rose-Hulman Institute of Technology in which Professor Mottel outlined a number of experiments which give rise to various differential equation models, $dy/dt = -ky^n$, $y(0) = y_0$, of order n to model kinetics. These included sublimation of CO_2 and evaporation of acetone in various shaped vessels (petri dish or funnel) as well as traditional zeroth, first, and second order reaction kinetics in chemistry. Mathematics faculty are reading *and writing* more widely, searching for examples for their students and seeking places where they can direct their students to witness mathematics in use and where their students can reach out and touch an application of mathematics.

How should science education reforms affect mathematics instruction (and vice-versa)?

The collection of real data in the field and its appropriate use to motivate and affirm a mathematical model is one example of how reform in science education can affect mathematics instruction. This data is often used to permit the student to assemble a personal model of some phenomenon. Consider the example of constructivism in science education offered by David T. Crowder (Faculty in Science Education at University of Nevada-Reno) in his article, "Cooperating with Constructivism," which appeared in the September/October 1999 issue of *the Journal of College Science Teaching*. The five main principles of constructivism, according to J. and M. Brooks in their text, *The Case for a Constructivism Classroom*, published in 1993, by the Association for Supervision and Curriculum Development in Alexandria, VA are "(1) Use the problem's relevance to students in instruction. (2) Structure learning around primary concepts. (3) Value students' points of view. (4) Adapt curriculum to address students' suppositions. (5) Assess students learning in the context of teaching." As an aside, on (3) — valuing students' points of view, how many of us have heard (or even engendered an atmosphere in our class which would permit) "well, it seems to me ...?"

Crowder has his students measuring the slope and velocity of bowling balls rolling down constructed ramps at the local bowling alley. Making reform in science education is relating science to activities with which students are familiar. This is not just in physics-for-poets classes, but rather in main line science courses and introductory engineering design classes offered by many engineering curricula to foster and keep engineering students interested in the field. It is meeting students where they are and developing them to be active learners. Some in mathematics instruction are doing just that in modest ways. An example of this is the team of Bruce Pollack-Johnson and Audrey Borchardt of Villanova University. In their newly designed business calculus course they require student-generated projects, i.e., students decide what they want to study and then build their own project using the mathematics at hand.

How is mathematics effectively integrated into the undergraduate curriculum?

Frankly, we do not believe mathematics is effectively integrated in the undergraduate curriculum. If it were, physics professors would not have to reinvent Fourier series for the students in optics, because instructors would encourage students to remember this from their work in separation of variables strategies in solving partial differential equations. Chemistry textbooks would not hide the facts of the differential equations describing chemical kinetics in the appendix; physics instructors would not be restricted to only the symmetric cases and to encouraging students to select the right boxed formula from their algorithmic physics texts; and engineering course work would not plod through old graphical approaches when students could use spreadsheets or computer algebra systems they learned to use effectively in mathematics courses.

There are probably several reasons for this lack of integration of mathematics into science and engineering curricula—and we believe it applies even more to the integration of mathematics into the social sciences. First, we have not done the best job in preparing students for the rough and tumble world of applications of the mathematics. Sure, our students can manipulate a bit, they can do the problem if they know the section of the book from which it comes, and they can push a bit, but not too far, with analysis. At times we have used different terminology, e.g., moment about a point and torque are used in engineering and physics and we need to tell our students they are the same when we first introduce the cross product to define torque. However, students cannot bring their mathematical tool kit to the problem when it is out of the context of a familiar framework. This may be because they are not truly familiar with what they have in their tool kit. More likely, it is because they have not practiced their mathematics out of the context in which they learned it, i.e., simple, one-step applications, as opposed to a timely use in the middle of a complicated situation.

Second, the way (order, logic, notation, motivation, etc.) we did the mathematics needed by the engineering and science faculty may not be suitable for their needs. It may be too general and thus need more refinement or it may not be encompassing enough and thus need expansion. Or it could just be that the professor of science or engineering who needs the mathematics we teach just does not like the way we did it and the students need to see it from the new discipline's point of view. Most often this is done without feedback or consultation with the mathematics faculty.

The way to integrate mathematics successfully is through dialogue with all interested parties. We should talk about what mathematics we do and *how* we do that mathematics. Throw in a healthy dose of *why* as well. Certainly, support and release-time help, but one-on-one conversation can kindle a great deal. Visiting classes can help, and using each other's texts as source material (e.g., take your data set from the physics lab manual, your current students' lab reports, or the chemistry text book). Faculty "in the trenches" have to be comfortable in the other discipline, not totally versed, but comfortable, and confident their new-found colleague can bail them out when they get in over their heads—which will happen! All the programs, all the funding, all the initiatives can work only if faculty will but talk to each other. They will do that if they believe there is gain for students and for them professionally. The external forces and actions exist to support faculty dialogue and motion—and talk and action are out there more than ever. Thus we are hopeful.

When should calculus be taught and what other courses are needed?

Perhaps calculus should never be taught! Perhaps what is needed is a modeling course in which rates of change and accumulation are the themes and perhaps we should call it "Rates and Accumulation with Application." This certainly sounds more inviting than Calculus, with the big C. We need to know the clientele, their background, their anxieties, their desires, and their goals in learning this mathematics. Faculty in other disciplines may not actually want to use anything we offer, they may just want to show it to their students. It is almost certain that these non-mathematics folks do not want their students to take our mathematics courses so their students can see how mathematicians think, or so they can learn to think like mathematicians. They have their own objectives in requiring our mathematics.

In the 1970s we conducted workshops on microcomputer models for life sciences. We used BASIC as the language of simulation, instruction, and presentation to both high school and college life science faculty. We never used the words "differential equation," but always used the words "rate equation." Indeed, we used a simple Euler step method to say "change" and "update" to our new value from the old value. We would change our step size and note the better approximation because the plots would look more reasonable, not because of some epsilonics, but because of our understanding of the modeled phenomena. We relate this here because at times we can truly scare off our colleagues with all sorts of high-powered mathematical or technical terms, e.g., first-order, non-linear, ordinary differential equations. Think about "first-order, non-linear, ordinary differential equations" and then say it out loud. Now say, "rate equations." Then get on with the exciting part for the scientist, modeling and "what if"-ing. This is what the client wants, not an opening chapter on classification of differential equations.

How is on-going involvement of the partner disciplines maintained?

At some expense to both parties! Do not deceive yourself. It takes energy and time and these cost money. In each field new pedagogies are emerging, and in some cases in different ways. Cooperative learning in a mathematics class where one is trying to discover an underlying principle might be very different than in the class on truss building. Expectations are changing in each field. Data analysis in one field can be viewed quite differently in another. For example, traditionally in chemistry and chemical engineering one linearizes the functional model, usually by logging the data, and then fits a straight line. However, in calculus optimization applications one may directly fit modeling functions to data through a minimization of least square sums, thus avoiding the entire activity of linearizing and plotting logged data to see if the transformed model fits a straight line and then backing out the parameters from eyeballed slope and intercept information.

There are practical considerations too. Consider how one could benefit from a colleague's professional society meetings. What would you go to at such a meeting if you had never been there before? And who would pay your way? Approach your department chair and say you need extra money for the other discipline's meeting where you want to go and visit and listen, not even present a paper. See what happens.

Conclusion

We believe there is hope for creating an interdisciplinary culture in undergraduate education. There is energy and light in this area already. It is appropriate to move in this direction at many schools. Faculty members want to join with other faculty to learn more about other disciplinary views and methods and to share this new interdisciplinary paradigm with students. As a society of educators we can offer support, both financially and temporally, and we can lead ourselves, but we have to have patience and prudence in our efforts. For to push too hard creates some backlash, certainly some resistance, and we need to understand the sensitive human interaction which takes place along with the intermingling of science, engineering, and mathematics. We are confident the future will be one that reflects the broad view offered by interdisciplinary approaches.

Interdisciplinary Culture—a Result not a Goal

Gary Krahn
United States Military Academy

Abstract. Interdisciplinary activities are ideal for nurturing the skills of reasoning. This allows the discipline of mathematics to become a thread throughout various academic disciplines. This note addresses why the process of mathematics education requires a continuum of double movements. The first movement often departs the environment typically associated with mathematics and enters other academic disciplines linked by problems and concepts.

Introduction

John Dewey in his classic book *How to Think* grapples with the challenge of defining the process of reasoning. Dewey states that *reasoning* is the recognition of relations of interdependence between considerations previously unorganized and disconnected. Developing the process of reasoning in students requires practice in the "recognition of relations of interdependence" through the analysis of problems. The literature on learning tells us that the best types of problems to engage students are those they deem most relevant, that is "real world" problems that people encounter. Such problems are rarely confined to the artificial boundaries of academic disciplines. For example, an apparently simple problem to arrive at a model for census data collection can simultaneously involve mathematics, social science, geography, and law. Thus, interdisciplinary activities are an ideal mechanism to develop students' reasoning processes.

Mathematics is the ideal base from which to develop interdisciplinary activities because mathematics includes the art of reasoning, the science of measurement, and the language of science. This definition is a gentle reminder that mathematics is not tethered to a specific set of courses or subjects, but rather that mathematics is a process. This process seems pleasant; however, a classroom journey to a lifetime of learning and discovery is extremely arduous. If the role of mathematics education is not well defined, this journey will be even more difficult. Mathematics reform is about redefining the role of mathematics in education. Mathematics must be focused more on the process of reasoning than on the foundation of knowledge and information.

Information vs. Wisdom

Thanks to technology and the development of the internet, information is ubiquitous. Giving students more information is equivalent to giving a drowning person a cup of water. In this information age, education is valuable if it provides an individual with the ability to think and reason effectively and efficiently, that is, the ability to sort and evaluate the plethora of available information. Therefore, to prepare today's students to function effectively in this information-rich environment, mathematics education must be rooted in developing reasoning ability.

Often the applications in core mathematics are aligned with topics and not necessarily with the goal to enhance the progression of reasoning skills. Applications should allow the student to associate their knowledge with their experiences in order to learn how to solve new and complex problems. Students must become comfortable moving back and forth between facts and ideas if they are to become truly educated rather than simply receivers of information. "Information is knowledge which is merely acquired and stored up; wisdom is knowledge operating in the direction of powers to the better living of life" (Dewey, 52). Ideal instruments for the development of wisdom are those "real world" problems that span different disciplines and departments. Thus, the "Interdisciplinary Lively Applications Projects" (ILAPs) a project of INTERMATH are an effective way to begin de-compartmentalizing traditional mathematical topics and to traverse course boundaries.

Reasoning

This broad responsibility to develop reasoning skills makes it very difficult to reach a consensus on which specific practices should be part of a mathematics program. Exposure to the art of reasoning, however, through the process of induction and deduction must be a component of mathematics education, The mathematics instructor must create learning opportunities where students move from *puzzling* data to a suggested meaning (i.e., induction). But just as importantly, we must require the student to move from the suggested meaning back to the data (i.e., deduction). In essence, a complete thought is a recursive process that involves both induction and deduction. The process of mathematics education requires a continuum of double movements. The first movement often departs the environment typically associated with mathematics and enters other academic disciplines linked by problems and concepts.

The inductive process requires the student to gain meaning from interdisciplinary perspectives. Data is the material of reflection and thinking. If this meaning is substantiated, the facts become evidence for new ideas. This meaning supplies a schema to understand the data more carefully. The second movement returns the student back to traditional mathematics as the schema is validated by reexamining the data with established principles and procedures. Deduction allows the student to seek additional observation to develop more powerful analyses through specialization. This return trip to the traditional mathematical environment becomes the starting point for another excursion into other interdisciplinary activities. The double movements of induction–deduction, expanding–contracting, generalization–specialization, interdisciplinary–disciplinary reflect the processes we want students to assimilate. A journey outside the discipline requires an interdisciplinary environment. Our experience shows that the interdisciplinary expedition is demanding for both the student and the teacher.

For example, in an introductory mathematics course we often introduce the topic of logic with *propositions* and *connectors* such as "or", "and", etc, using truth tables. Tautologies are quickly introduced and discovered. The stage is now set to reveal how logic can become the foundation for circuit designs that govern every electrical device imaginable. These excursions into an interdisciplinary activity, however, should never be aborted because of a lack of knowing the applications. Our educators should be encouraged to dialogue with members from other departments. For this example, a visit to the electrical engineering department would generate material to encourage students to discern the engineering importance of forming equivalent statements that minimize the number of operators. Other classes may explore the application of logic and valid arguments in law. Using activities from other disciplines, discoveries and return trips are made to the fundamental constructs of logic.

It is not the case that interdisciplinary activities always represent advancement while the creation of discipline-specific activities is degenerative. It is the coordinated movement between discipline-specific and interdisciplinary associations that promotes effective education. We must recognize that in academia, interdisciplinary activities and specialization must work in concert to complete the learning cycle. Education is a process that must involve both interdisciplinary activities and discipline-specific knowledge.

Ultimately, we want to simultaneously unleash and coordinate the development of our students, faculty, and society. Interestingly, unleashing and coordinating are contradictory actions as are the concepts of unification (interdisciplinary) and specialization. Our challenge is to find ways to create a learning environment where the dynamics of dismantling academic boundaries while focusing on specific subject matter advance education and research.

Should calculus be taught?

Historically, in mathematics we have immersed students in detail and loaded them with disconnected facts. Simultaneously, the classroom discussions are rapid and directed. In the classroom once a student makes a conjecture, if it is correct, the teacher accepts it; if it is false, it is rejected. The teacher, who assumes responsibility for the students' intellectual development, usually amplifies the idea. Reform mathematics, however, encourages the student to form generalizations of the facts and then requires the student to examine the implication of these generalizations. This student-centered approach can be effective only if students can relate to the problem, and real world problems tend to be those with most relevance for students, and those problems cross disciplinary boundaries. Hence, the process of induction and deduction flourishes in an interdisciplinary environment—providing motivation, meaning, and perspective to the reasoning process.

The standard mathematics courses, such as calculus, linear algebra, probability, statistics that are immersed in topics, do not align very well with the concept of reform mathematics. Ideally, a traditional course in calculus should never be taught. Traditional courses by their very nature emphasize topics over process. A core program should furnish the students opportunities at transforming a problem statement into a mathematical model, conjecturing solutions, selecting or developing the appropriate mathematics, examining the analysis, and continuing to transform the conjecture into a solution. A core program should be part of a curriculum that is crafted to promote the reasoning process rather than carefully visiting a set of topics.

Conclusion

There is no single program or person that will propel mathematics education to great success in the next millennium. The foundation for change in mathematics education, however, must be based upon the development of the processes of reasoning. The result of this change will inevitably be an interdisciplinary culture. Interdisciplinary activities are a result of mathematics reform, not a goal of reform.

We live in a time of profound change. Changing technologies require us to continuously assess the appropriate role of human capabilities in the process of problem solving. The next twenty years promise to be an eventful journey in education.

References

1. John Dewey 1991, *How We Think*, Buffalo: Prometheus Books (Originally published: D.C. Heath, Lexington, MA, 1910)
2. David Hume 1955, *An Inquiry Concerning Human Understanding,* edited with an introduction by Charles W. Hendel *Imprint* New York, The Liberal Arts Press.

Mathematics for Use Across the Curriculum
A Tool, a Language or What?

Robert G. Fuller
University of Nebraska – Lincoln

Abstract. The metaphors that faculty use to describe the process of mathematics can have a significant impact on the role of mathematics in the classroom. This note examines the residue that might linger in the hearts and minds of students when faculty state that mathematics is "*a tool, a language,* or *the manipulation part*" of the problem solving process. We must be careful. Metaphors depicting mathematics are often equivalent to saying that a spoon, knife, and a fork is dinner. Therefore, what metaphor should we use?

Introduction

What metaphor shall we use to most adequately convey the goals and intentions of the mathematics across the curriculum movement?

The metaphor we adopt for describing the use of mathematics in other disciplines is an essential part of communicating understanding to faculty and students. The metaphor we select is rooted in our understanding of what it means to know and use mathematics. This understanding reflects our mathematical epistemology; i.e., what we believe it means to know mathematics. Our epistemology, properly formulated, includes our understanding of both mathematics and the thinking of the people who use mathematics.

Let us begin by examining three different ways to talk about mathematics in the classroom. Given that most students in introductory mathematics courses have no intention of becoming professional mathematicians, the way we describe mathematics can have a significant influence on their attitude and their future uses of mathematics.

"The rest is just mathematics...

What understanding of mathematics do you think is presented to the students by the phrase, "The rest is just mathematics..."?

This phrase is sometimes heard in science and engineering courses at some point in a problem solving exercise. For example, a heuristic that has been used in some physics courses is ***EDPIC***, an extension of the problem-solving strategies advocated by Fuller (1982) and Reif (1976). The letters stand for the phases of the problem solving process that is used.

Exploration: Examine a physical system and collect some experimental data.

Description: List explicitly the given and desired information. Make any necessary assumptions. Draw a diagram of the situation. (The result of this step should be a clear formulation of the problem.)

*P*lanning: Select the basic relations pertinent for solving the problem and outline how they are to be used. (The result of this step should be a specific plan for finding the solution. Most physics problems involve mathematical reasoning and model building.)

*I*mplementation: Execute the preceding plan by doing the necessary calculations. (The result of this step should be a solution of the problem.)

*C*hecking: Check that each of the preceding steps was valid and that the final answer makes sense. (The result of this step should be a trustworthy solution of the problem.) If time permits, refine the model.

Typically, the students do an interactive video application, hands-on activity, or microcomputer-based laboratory data gathering in the **exploration** activity. This provides them some feeling and ownership for the data, as well as experience with the actual physics principles. The **description** begins with a drawing or diagram, such as a vector force diagram, free-body diagram, or sketchy graph. The relevant physical principles such as Newton's law of motion, or the ideal gas law, enter as the **planning**. At this point, typically, the students have a reasonably difficult mathematical problem to handle. Sometimes handling the equation involves graphing, sometimes solving a differential equation. In either case, a computer algebra system (CAS) is an excellent platform for **implementing** the mathematical analysis of the model. Finally, the students derive the answer or result with mathematical modeling. The result is **checked** against common sense, well-known physics results, the experiment, or the results obtained by other students or groups in the class.

When such a heuristic is used in a physics classroom, it is not unusual to hear the instructor work through the first three parts of the strategy, E-P-D, and then say, "the rest is just mathematics…". From there the instructor may, or may not, go through the rest of the problem solving strategy in detail.

Please take a few moments to examine the implication of that phrase and what it may show about the mental model the instructor is conveying to the students about mathematics.

In this instance, the instructor seems to be conveying that, from the point-of-view of the instructor, the difficult and challenging parts of the exercise are completed by the time the instructor claims to use mathematics. In this setting, mathematics is assumed to be a simple algorithm the problem-solver can use to obtain a correct answer. In this context, mathematics is seen as subservient to the larger and more interesting problem-solving strategy being used in physics. On the other hand it points to the usefulness of mathematics in reaching a satisfactory answer to the problem.

Unfortunately this reference to "just mathematics" neglects the mathematical reasoning that is inherent in most of the laws and principles of physics. The predictive nature of physical laws is based on their mathematical structure.

Nevertheless, since the setting up of a problem is still a long, long way from a useful solution, maybe the mathematics across the curriculum movement should adopt as its slogan, "the rest is just mathematics…"?

Mathematics is a tool to…

What understanding of mathematics do you think is presented to the students by the expression, "Mathematics is a tool to …" ?

A common way of talking about mathematics in science and engineering courses is to view mathematics as a tool. In these contexts, the task at hand is typically a problem to solve to obtain a result. The result requires the doing of a variety of operations for which particular tools are especially useful or needed. A difficult task is made much easier by having the proper tool. Professionals nearly always have the correct tool available for their use. Hence, in the context of science and engineering as professions, mathematics is presented as a tool to be used for the successful completion of the task.

Please take a few moments to examine the concept of a "tool" and what it may reveal about the mental model the scientist, or engineer, is conveying to others about mathematics. Perhaps, in this instance, the professional is showing that from this point-of-view the task can best be accomplished by the use of mathematics. In this setting, mathematics is a device, or technique, the problem-solver can use to obtain a correct answer. Mathematics is seen as serving the larger and more interesting problem solving strategy being used in this course. While at the same time it shows the importance of mathematics is reaching a satisfactory answer to the problem.

It can be argued that the "tool" metaphor for mathematics is rooted in a behaviorist view of knowledge, a view that sees problem solving as a process to obtain a correct answer. This view of knowledge sees the correct answer as predetermined by the problem. The process of figuring out the answer will have no influence on what the correct answer really is. In this view of knowledge the tool influences neither the process nor the answer. This view of knowledge sees that the tool and its properties are completely independent of the task.

To what degree does the tool metaphor reflect the understanding of mathematics by scientists? Or engineers? Or mathematicians? In what ways is it a useful metaphor for the mathematics across the curriculum movement?

"Mathematics is a language for…"

What understanding of mathematics do you think is presented to the students by the expression, "Mathematics is a language for …"?

Physics is sometimes described as the building of mental models to explain and predict the behavior of natural systems. Such models are always linked to the measurement of some physical observables of the system. The models may have exotic properties like eleven dimensions of a string, but they are based on physical observables. The models of physics are most productive when they use the language of mathematics to convey their symmetry and structure. In fact, there are examples in the history of physics where the exploration of a physical system led to the need for new mathematics. On the other hand, beautiful mathematical systems have been developed without regard to any requirement to be an expression of the physical universe, only to be discovered, perhaps years later, to be just exactly what the physicists needed to help them complete an analysis of some natural phenomena.

The mathematics-as-a-language metaphor seems to have its roots in a constructivist view of knowledge. In this view, as a problem is being solved, the scientist, or engineer, is constructing a framework for describing and explaining the problem. This constructivist epistemology sees the interaction between the problem-solver and the problem as an essential feature of the process. In such a dynamic process the communication of the activity is carried out among human beings via language. If the results of the problem-solving process are going to be given by a numerical, or mathematical, result, then a mode of communication based on the principles of mathematics will be required. The grammar and vocabulary of mathematics will become essential aspects of the problem-solving process. These attributes of mathematics will interact with the mental models being developed by the problem solver to shape and transform the mental activity of the problem-solver and the nature of the answer being sought to the problem. In this view of mathematics the dynamic interaction between the problem solver and the problem occurs through the intermediary of a mathematical language which helps to provide structure to the problem-solving process and the mental models being used. In addition, mathematics becomes a lively part of problem solving and is altered itself by the process. Such a dynamic view of problem solving requires that the subjects and objects of the process are dynamic and open to transformation during the process. Clearly, real-life problem-solving experiences have such characteristics. As people seek to solve problems in their lives, the information and the context of the problems get changed as the person interacts more and more with the problems at hand. The dialogue between the problem-solver and the problem also changes the language used to communicate the problem-solving process.

In some science and engineering courses, students will hear mathematics discussed as the language used in the problem-solving process of the discipline. While this seems to be a particularly rich metaphor for the use of mathematics it is infrequently discussed in any detail in a science, or engineering, course.

Perhaps, filling out the details of such a metaphor is a useful task for the people who are participating in mathematics across the curriculum activities.

Ultimately, the success, or failure, of the mathematics across the curriculum movement may depend upon the metaphors that we use to describe mathematics. It is an important task that demands our attention. What metaphors shall we choose?

References

Fuller, R.G. 1982. "Solving physics problems—How do we do it?" *Physics Today* 35(9), 43–45.

Reif, F. J. H. Larkin, and G. C. Brackett 1976. "Teaching general learning and problem-solving skills," *American Journal of Physics* 44, 212–217.

How Do We Describe an Interdisciplinary Curriculum?

Thomas R. Berger
Colby College

"Would you tell me, please, which way I ought to go from here?"
"That depends a good deal on where you want to get to," said the cat.
"I don't much care where ..." said Alice.
"Then it doesn't matter which way you go," said the cat.

Alice's Adventures in Wonderland, Chapter 7

Abstract. What are the expected outcomes of an undergraduate education? This note reflects upon some of the skills and attributes that have been talked about in CUPM meetings. As with Alice in Wonderland, if we don't know what students need to experience, then it seems not to matter what the curriculum should be. We look at some of the content goals for the curriculum and goals for the educational process.

Introduction

The Committee on the Undergraduate Program in Mathematics (CUPM) is addressing the various majors in the mathematical sciences. About 1970 [1] CUPM released documents describing the major in terms of some courses and their mathematical content. By the early 1980s [2] this task was more difficult so the resulting document is more ambiguous. A more recent effort in the early 1990s [3] was not conclusive since the scene was rapidly changing and the situation very diverse. In 1995 [4] CUPM took another approach. They looked at successful programs and presented these as models of good curricula. There is a growing feeling that these case studies do not go far enough. It is time again to try to describe a curriculum.

During this same thirty-year period attitudes changed about curriculum based upon broad experience with massive numbers of mathematics students and with research on teaching and learning. There is a general feeling that we should describe education in terms of expected outcomes. No one has exactly figured out how to do this yet. In other words, the major should have goals for the students and the program. If those goals can be articulated, then they can be combined with information about a school's students, the faculty teaching those students, the types of careers for graduates, and the environment of the college or university. From these factors a program can be planned that fulfills the goals and serves the students, institution, and faculty.

Probably even greater have been the outside forces that act upon curriculum decisions. Just to list a few:

- First and foremost, technology has affected the tools we have to teach. It has become a demand from employers. We ignore it at the peril of our students.

- Second, a wide range of teaching styles have been tried and discussed. Much research has been done on student learning. The results indicate that these different teaching styles may have much to recommend them.
- Third, our majors are no longer prepared primarily for entry into graduate study in mathematics. Many graduates go on to graduate and professional schools, but mainly in disciplines other than mathematics. The vast majority of graduates enter the workforce directly. We can single out a few places (for example, teaching, finance, and engineering) where it might be possible to delineate what mathematics students should know. [5] Except possibly teaching, these few areas still attract less than one-half of the majors. Diversity is the most salient feature of careers of recent undergraduates in the mathematical sciences.
- Fourth, employers seek mathematics connected with other areas of knowledge. Students should be able to translate the mathematics they know into the situations they will encounter on the job. This exerts pressure on faculties to develop courses in "relevant mathematics" when sometimes we don't even know what the word "relevant" should mean.
- Fifth, because so many students enter the job market, additional skills are expected. The students may need technical writing skills, practice in interpersonal relationships in a technical working group, experience studying with students in a variety of disciplines, and other very non-traditional skills.
- Sixth, a discovery that the success of students in a major has a great deal to do with the context in which learning takes place. The respect students receive, the care taken for their welfare, and the pleasure they take in their studies are important factors. The excitement and pleasure faculty take in their courses is another important factor.

With all this in mind, how can we specify our expectations of a major in the mathematical sciences in such a way that a department might plan a program? This paper presents some preliminary ideas that have come to CUPM. First, how might a description of a curriculum be given if it describes what students should know at the end of two years? Second, if process skills are important, which ones do we describe and how do we do that? The paper mostly just raises questions and offers a few tentative examples.

Goals for Content

How might a description of expectations look? No one really knows, but many are trying. This paper makes a modest start by using calculus as an example. For the curriculum in the mathematical sciences majors, calculus is probably the most described and taught of all courses. Describing our expectations ought to be easy. It is not.

In the past it was expected that students should have completed a course in the calculus at the end of the first two years for a mathematical sciences major. A traditional curriculum would discuss the usual or minimum coverage of topics. Turning this around, we ask ourselves the question:

"What knowledge does the average student carry out of this course that is durable or will be heavily exercised in future coursework?"

At the most basic level of curriculum description, what skills do we expect of students? Do we expect students to know differentiation and integration formulas and techniques? Experience and research shows we rarely attain the traditional goals we set for skills even with very good students. Now computer algebra systems (Maple, Mathematica, and Derive) provide powerful tools to simplify traditional differentiation and integration problems. If so, do we want the same level of competence that traditional curricula prescribe? A perusal of textbooks shows that these matters are not resolved. Essentially all texts claim to be fully *calculus reformed*. Some texts place heavy emphasis on traditional skill development (including some of the most widely used texts). Others downplay this role while increasing emphasis on conceptual understanding.

It is in this context that we must describe our expectations for students of the calculus. (Others have discussed calculus [6,7] and the content of the first two years. [8]) Here then are some possible expectations of our students for calculus.

1. Every mathematical sciences major should have an intense calculus experience lasting at least one year.

   A sample attempting to describe a student by level of competence is given next. This standard should be applied one or more years after the course is completed. For curricular purposes, it provides a basis for discussion of the calculus course and how it fits into the mathematical program. Let's try to describe computational levels of achievement.

2. A student completing a calculus course should achieve computational proficiency in the course. This might be measured as follows.

   a. Basic Proficiency: The student is presented with a problem typical of the course and its solution. The student can read and understand both the problem and its solution. The student can describe each step leading to the solution.

   c. Proficient: The student has basic proficiency and can describe the solution by explaining the methods used, why they were used, and how they were used. The student exhibits some understanding of the connection of this problem with general knowledge in the subject.

   d. Skilled: The student is proficient and can frequently supply solutions to such problems without being presented with them, especially if given a little warning and time.

   e. Highly Skilled: The student is skilled and can solve some complex problems requiring significant symbolic computation (either manually or with the help of a computer algebra system or both). The student can take a page of such computation and discuss and modify it in interaction with another highly skilled person.

   This approach to a curriculum is sensitive to the kinds of problems posed for students. A traditional curriculum would focus on problems now familiar to us. An interdisciplinary curriculum would emphasize problems rooted in a variety of disciplinary contexts. For today's students, problems should probably contain a mix of applied and pure. Understanding how the calculus tells us about the economic, social, living, and physical world around us broadens students' understanding of the mathematics itself and provides a basis for making mathematics useful.

   The levels and descriptions above are similar to graduation standards being written around the country for high school students. For a school student the description might refer to an algebra, rather than a calculus course. Recently I spent a day looking at statewide test results in order to obtain a community consensus for performance levels described as above. The goal for skill level for the course would be that all students achieve, at least, Basic Proficiency with some students becoming Highly Skilled. This rather elaborate style can be continued through each of the following types of knowledge. I'll be a bit briefer.

   Skill development is only part of understanding calculus. A curriculum should provide a foundation for a conceptual understanding. Stating measures of achievement in conceptual understanding is more difficult since our tests tend to only indirectly measure this quality.

3. The student should be proficient in the basic principles of the calculus.

   a. A student should be able to describe the derivative by means of examples with tables, graphs, and formulae, including interpretations as both slope and instantaneous rate of change. The student should be able to move comfortably among the various representations. The student should have some notion how the derivative is approximated by the difference quotient. Depending upon the level of the course and expectations of the program, students may be able to relate these concepts with their mathematical explication including connections with the difference quotient and limits. If a course has a theory component, students should be able to explain connections with definitions and simple theorems. They should be aware that the subject depends upon fundamental properties of the real number system.

b. A student should be able to describe the integral by means of examples with tables, graphs, and formulas, including interpretations as both area and total change. The student should be able to move comfortably among the various representations. The student should have some notion how the integral is approximated by sums of areas of rectangles. Depending upon the level of the course and expectations of the program, students may be able to relate these concepts with their mathematical explication including connections with Riemann sums and limits. If a course has a theory component, students should be able to explain connections with definitions and simple theorems. They should be aware that the subject depends upon fundamental properties of the real number system.

Are standards like these for content appropriate?

A curriculum should provide a basis for discussion within a department for a curriculum appropriate to the students, department, and institution. Because these factors can vary so much, it isn't possible to list every topic. Enough information should be provided to guide a department to make wise decisions compatible with those being made at other institutions.

What other content expectations should be placed upon students for the first two years of a mathematical science major?

(In all programs and majors in the mathematical sciences.)

A. If the student is specializing in an area like statistics or computer science, should we place an expectation of completing an intensive yearlong course in the specialty?

B. Do we expect every major to have experience with linear algebra in the first two years?

C. Do we expect every major to have experience with differential equations in the first two years?

D. Do we expect every major to have experience with infinite sequences and series in the first two years?

E. Do we expect every major to have experience in discrete mathematics in the first two years? (Topics from computer science, combinatorics, optimization, mathematical modeling, etc.)

F. Do we expect statistical learning for all mathematical science majors in the first two years?

G. For pure mathematics majors, do we expect introductory experience with proofs at an appropriate level?

H. Do we expect that students will also develop interdisciplinary knowledge in appropriate courses? If so, how should this be described so as to serve a diverse student body?

Students will be exposed to this variety of mathematics in a sequence of courses. Should there be summative demands on the curriculum so that students pull together the knowledge they are acquiring from various courses? Issues of content are ones we claim to understand. Can we find sufficient consensus to write content standards for the first two years?

Goals for Process

The world expects our students to have skills usually not explicitly taught in school. Might it not be time to make some of this explicit? Let me state some particular skills so that you can react to them. These types of skills involve general competencies applied particularly to the mathematical enterprise. Frequently in English courses students learn how to use the campus library. But when asked to do so in a mathematics course, the students are often unable to perform the simplest tasks. That is, our assumption that "English can take care of this issue." is not valid. Knowledge does not transfer quite so easily as we hope. So we may want to help our students learn process skills in the context they will use them within our discipline.

Here are some possible process skills for the first two years. Are there ones on the list that are optional? Are there missing ones that are crucial?

1. Learning skills.
   a. Students should be able to join a team and learn the basic principles of a new topic from a given body of text or reference material.
   b. Students should, on their own, be able to learn the basic principles of a new topic from a given body of text or reference material.
   c. Students should be able to acquire the basic principles of a new topic from a lecture at the appropriate level.
2. Resource Skills
   a. Given a specific mathematical topic, students should be able to
      i. find resources at the appropriate level on the given topic in a library,
      ii. find resources on the Web and know how to check for validity and accuracy, and
      iii. find resources within the community.
   b. Students should be able to appropriately organize the results of research for the task at hand.
   c. Students should know about standards for plagiarism, bibliographic style, and presentation style within their area of concentration.
3. Communications Skills
   a. Students should be able to write solutions to problems in a way that communicates to a general mathematical audience.
   b. Students should have experience in writing extended reports on mathematics.
   c. Students should be able to orally present mathematics of the appropriate level to a group of peers.
4. Working Skills
   a. Students should be able to learn to use computer tools and have some knowledge of their own learning curve for such tools (i.e. how long will it take to learn and what level of effort must be invested?).
   b. Students should work effectively as a member of a team on an extended mathematical problem.
   c. Students should have experience working on an interdisciplinary team on an interdisciplinary problem.
5. Problem Solving Skills
   a. Students should have experience in working on extended mathematical problems?
   b. Students should understand problem-solving processes and be able to articulate and apply these processes?

For many years, in several engineering and science disciplines, students have been given ill-posed, large-scale problems and asked to form teams to attack these problems. Sometimes courses on mathematical problem-solving address issues of this kind. These experiences are valuable for students seeking jobs right after graduation. Many students who continue to graduate study also testify to the value of such experience. Might there then be process standards that are summative in this sense?

6. Summative skills

   Students should have experience working on a team and bringing together most of their process skills and much of their knowledge. The experience should involve problem clarification, resource gathering, problem solving, application of mathematics, use of appropriate technological tools, report generation, and written and oral communication of results.

In no sense are the above skills being proposed as essential to a curriculum. However, they are given in order to pose the question:

Should curriculum standards for mathematics describe process skills?

In other words, should we look at our entire curriculum to see that a major does acquire a certain minimum of process skills?

Other Goals

I have only suggested content descriptions for some of calculus and for a few process skills. There are many other aspects of a curriculum. For example:

1. How serious are we about an interdisciplinary culture in our curriculum? Are we serious enough to ask that applications of a subject be more than peripheral? Should this topic be addressed in a curriculum description?
2. Students learn in a changing environment. In other words, the mode of learning is itself an integral part of the curriculum. How should a curriculum description address this issue?
3. Technology has been implicit in many of the suggestions above. Does the issue need explicit mention in a curriculum?

Summary Comments

The book "Models that Work" [4] makes interesting reading. Frequently courses are mentioned but never quite described in terms of content. A successful curriculum depends very heavily upon the context for learning and its appropriateness to the faculty, students, and university. In trying to understand why the models are effective, one wants to read studies like: [9] "Talking About Leaving," a study trying to understand why bright students leave mathematics, science, and engineering majors. [10] "What Matters in College?," a summary of many years of research by Astin's team isolating factors that influence student success in college, and [11], "How College Affects Students," a summary of twenty years of research on all facets of the college experience. Since the success of a curriculum includes the context in which courses are delivered, how should this aspect of the curriculum be discussed in a curriculum document?

In "Commentary on a General Curriculum in Mathematics For Colleges" (See [1], Volume I, page 33), CUPM describes majors in the mathematical sciences by listing courses and their content. At the entry level, topics are described in some detail including suggested time for presentation. These types of descriptions focus attention on coverage rather than understanding. A shift to descriptions of expected student outcomes might refocus attention on what students actually learn. Since experience with outcome descriptions is limited, the task is difficult. I've attempted to give some feeling for the kinds of descriptions currently being used.

Now that a vast majority of mathematical sciences majors enter the workforce immediately upon graduation, the process skills we address in the curriculum become more important. They provide the context in which the students do mathematics. Experience with process skills enables students to use their mathematics. I've attempted to mention a variety of skills that have been discussed or mentioned in CUPM meetings.

In order for a curriculum description to be useful to the larger mathematical community, we need a great deal of discussion and reaction. Your suggestions and reactions are welcome. As you write about these issues you might try to spend some time focusing on goals and expected outcomes for students. You can address email comments to CUPM at `cupm-curric@ams.org`.

References

1. Committee on the Undergraduate Program in Mathematics 1973. *A Compendium of CUPM Recommendations*. Volumes I and II, Mathematical Association of America.

2. Committee on the Undergraduate Program in Mathematics 1981. *Recommendations for a General Mathematical Sciences Program*, Mathematical Association of America.
3. Committee on the Undergraduate Program in Mathematics 1991. *The Undergraduate Major in the Mathematical Sciences*, Mathematical Association of America.
4. Alan C. Tucker, ed. 1995. *Models that Work*, Mathematical Association of America.
5. The evolving document on the *Mathematical Education of Teachers* is a description of the curriculum for future teachers of mathematics.
6. Anita Solow, ed. 1994. *Preparing for a New Calculus*, Mathematical Association of America.
7. A. Wayne Roberts 1996. *Calculus The Dynamics of Change*, Mathematical Association of America.
8. John A. Dossey 1998. *Confronting the Core Curriculum*, Mathematical Association of America.
9. Elaine Seymour and Nancy Hewitt 1994. *Talking About Leaving*, Bureau of Sociological Research, University of Colorado.
10. Alexander Astin 1997. *What Matters in College?,* Jossey-Bass.
11. Ernest T. Pascarella, Patrick T. Terenzini 1991. *How College Affects Students*, Jossey-Bass.

Technology Perspective

This section edited by Patrick J. Driscoll.

Technology has been both a driving force in curriculum reform and a source of much debate and discussion since the late 1980s when graphing calculators were introduced and computer algebra systems (CAS) became available on desktop computers. The ability of CAS to present dynamic visualization, compute numerical solutions and approximations, obtain closed form symbolic solutions, and perform rapid iterations has brought about many changes in the study of mathematics. In particular, we bear witness to an almost universal de-emphasis on the relative importance of students performing mechanical calculations in comparison to developing their deeper thinking skills: abstraction, mathematical representation, interpreting results, and communication. In short, technology has refocused curricula on the modeling process and away from the solution process. The resulting reduction of hand calculation skills is one major source of the controversy surrounding the use of technology, especially in light of the service role to other departments that some mathematics courses perform.

The authors in this section investigate various perspectives related to the impact of educational technologies within the modern mathematics curriculum, examining the difficulties involved with integrating these technologies, and the potential promise technology holds for leading further curricula reform.

L.G. de Pillis explores the issues of "Technology as an End" and "Technology as a Means," and the impact these perspectives have on curriculum development. The former is usually minimized or absent from the curriculum even though knowledge of it lessens the "black box" syndrome and helps the student to develop uses of technology as a tool. A fundamental reason for incorporating technology in our teaching is to prepare our students for the technological world into which they will graduate. Charlie Patton provides some insightful experiences with computer software, highlighting the shortcomings of software design to truly imbed the needs of technology-enhanced learners.

Joe Myers examines the issues associated with technology tradeoffs and the emergence of a new learning style among students who have grown up with technology — "Many students will ignore the texts on the shelf in front of them and instead surf to research and find information." What impact will this new learning style have on curricula? Frank Wattenberg underscores the importance of making experimentation the backbone of a student's mathematical experience, citing several examples of how this might be achieved using currently available technologies and focusing on the interplay between modeling and application.

The value of considering the student perspectives and expectations in curriculum design is discussed by Patrick Driscoll. He introduces a general framework for how educational technologies fit into mathematics curriculum development, illustrating how various relationships that exist between student and faculty perspectives have an impact. John Scharf continues the discussion in this vein, suggesting that the assessment methods we use to facilitate student learning must evolve as the curriculum evolves. Noting that students arrive in college more comfortable with change than most faculty members, and more willing to experiment with technology than most faculty members, he advocates that all students gain experience modeling realistic problems so that they "see the importance of mathematics as they learn it." Lee Zia highlights the institutional perspective, proposing a curriculum model that seeks to take advantage of the natural synergy of these three core mathematical subjects and their place within the broader context of second-year science, mathematics, engineering, and technology education. Wade Ellis discusses the anxiety and reluctance to address applied problem solving as a notable trend in the mathematics community, positing that the inertia of change in the educational landscape makes applied problem solving a necessity.

The authors all provide their perspective as to some of the benefits the mathematics community is able to realize through the use of technology:

- Visualization — graphics provides the first step in problem solving.
- Modeling and Demonstration — simulations of physical experiments can be carried out computationally and systems can be seen to evolve in real time.
- Discovery (experimentation) — elimination of time-consuming hand calculations, the ease of experimenting, and visualization capabilities invite student inquisitiveness leading to conceptual understanding and discovery.

On the other hand, the authors readily identify several of the principal detractors associated with technology that appear to limit its integration into current course design:

- Time investment — "start up time" to become competent and comfortable with a particular software or calculator, time to update to a new version, planning time to effectively incorporate technology into the curriculum.
- Ease of use — The practical issues of getting the equipment properly set up sometimes interferes with or discourages the use of technology in the classroom.
- Decrease of hand calculation skills — this is the major source of the controversy swirling about the use of technology in teaching.

The explosion in technology will continue to produce "tools" for learning and doing mathematics. The evolution of flexible, general-purpose tools will enable students to explore ideas beyond the confines of a text and prepare them for the open-ended experimentation and analysis required in the workplace. Charlie Patton suggests that as the computer fills the role of the textbook in the future, there are numerous challenges and opportunities on the horizon. They primarily fall into the areas of discovery, communications, and assessment.

The Machine is the Textbook

Charles M. Patton
MathTech Services

Abstract. Substantial advances in computational software and applications ported to handheld devices have greatly assisted in relaxing the bounds on the types of problems accessible to students in a mathematics curriculum. Be that as it may, we submit that very few of these programs have been developed with effective pedagogy and student learning central to their design. In this paper, we present several throughts on this shortcoming and suggest ways in which the current state of technology could be improved to better meet the needs of an ever-changing audience of learners.

Introduction

Be it cultural evolution or cultural degeneration, students are looking more and more to "the machine" (calculators, computers, the Internet, video, gameboys, etc.) for guidance, insight, and entertainment. Virtually all the math software available today, however, is built on an architectural plan laid down in the "textbook era". It is not at all obvious that any of it can meet the challenges, take advantage of the opportunities, or live up to the responsibilities which obtain as we leave the textbook behind.

These responsibilities and opportunities fall into roughly three areas whose general descriptions are as follows: the intellectual scaffolding necessary to allow students to explore, make connections, gain experience (DISCOVERY); the infrastructure necessary to allow for this most human of activities to facilitate collaboration among students—when appropriate, and allow students to present their work naturally and efficiently (COMMUNICATION); the system necessary to enable a student's act to become an artifact to be observed, recorded, reflected on, and ultimately judged (ASSESSMENT). Technology can be a major contributor or facilitator in each of these areas, but first the needs and expectations within these categories must be clearly exposed.

The potential for student discovery has been a driving force for the adoption of technology in the classroom at all levels, especially in the calculus. Graphs, tables, formulae, numeric experiments, integration, differentiation—all provide for an environment in which students can experience a range of calculus concepts for themselves, and in their own way. The advance of discovery tools for calculus, which was quite rapid in the late 1980s and early 1990s, has slowed down considerably in recent years. In part, this is because the tools available to advance the state of the art were designed first and foremost to get answers. Up to a point, you can use an answer-oriented tool to help with insight by "throttling down" the power of the answer engine, making it go more slowly—perhaps even step by step—but adding more and more sophisticated answer-finding capability is unlikely to add more insight power automatically.

PUFM

Roger Howe [1] in his review of *Knowing And Teaching Elementary Mathematics: Teachers' Understanding of Fundamental Mathematics in China and the United States* [2] highlights Ma's notion of a "Profound Understanding of Fundamental Mathematics" (PUFM) and while Ma's book targets primary and secondary school mathematics, Howe clearly believes that PUFM is just as relevant in college teaching.

One aspect of PUFM is the knowledge of the connections among and analogies between mathematical and physical concepts. So, for example, a teacher would be able to relate the notion of derivative with approximate local linearity, which builds on both the model of linear functions and interpolation, which in turn builds on the notions of functions, composition, and proportionality, which builds on addition, subtraction, multiplication, and division. In turn, higher dimensional (even infinite dimensional) versions of the derivative are built on this same notion.

The only way technology could help students discover these roots and draw mathematical nourishment from them would be to have these connections present and manifest at deep levels of the software itself. Although no software system in use meets this requirement today, several technologies, long in the works, were ushered in during 1998 and 1999 and hold great promise as a starting point for implementing a system with this kind of depth.

One of these is the MathML specification [3] which, among other things, provides an unambiguous, machine-readable grammar for describing the intended meaning of a mathematical expression, including its context, annotations, and links to other conceptual structures.

At the same time, work is progressing on laying out conceptual structures which could capture some of the richness of mathematical connections. In particular, the Math MetaData project [4] aims to create an extensible, inter-linked classification system that both incorporates the AMS Subject Classification and extends into the undergraduate and K-12 learning environment, while providing for appropriate concept, topic, subject, and class interconnections.

The dream is that as these two efforts progress, it will be possible for a student looking at the derivative of a composition of functions to ask of the system "is there anything like this I might have seen before?" and being shown the slope of the composition of two linear functions. Repeating the query with the new expression, the student might be shown that if A is proportional to B and B is proportional to C, then A is proportional to C.

Multiple, Multiple Representations

A recurring theme arising both in the NCTM Standards efforts and in the calculus reform efforts is that of linked multiple representations of mathematical concepts — often graphical, numeric, and symbolic. While these are all important and well represented in current technology, they don't go far enough. For example, a significant impediment to integrating mathematics and engineering instruction is simply the difference in representation of the underlying mathematics. Something as simple as Newton vs. Leibniz notation can derail joint understanding. So can the use of "i" vs "j".

Beyond that, there are interesting and useful representations which, because they were not in widespread use at the time of their introduction, have not been imbedded in today's software (for example, the nomograph representation [5]). However, in contrast to simply accepting this as a fixed operational characteristic, there is a growing subscription to the notion that it is the *viewer* who decides which representation is the most informative. This constitutes a quiet revolution, and has fostered the birth of efforts like the Educational Object Economy [6]. Combining the strengths of markup language and style sheet technologies with a widespread, unambiguous means of describing mathematical content, efforts such as EOE allow the proliferation of new ways to view old constructs.

Communication

Besides just replacing the textbook, technology provides opportunities to address issues that textbooks could not address effectively. One of those issues is communication, especially communication in the classroom, especially structured conversations for learning. Much of what transpires in the average classroom is structured conversation. It is structured both in form — who converses with whom and under what conditions — and in content. In the pure lecture-style class, any technology beyond what is already widespread is probably not very necessary. However, as the role of teacher evolves from being a "Sage on the Stage" to being a "Guide on the Side", from individual to collaborative efforts, from passive listening to active response, the logistical and administrative costs of the structured conversation rise significantly, and so does the opportunity for technology to make a difference. .

Supposing that the technology could provide for conversation structured in form (who converses with whom under what conditions) and in content ("live" mathematical objects, dialog and commentary), what could be done with such a system is yet to be determined.

There are many situations where a deeper insight about a mathematical process might be obtained by having each student focus on an individual local problem, while observing how their local solution contributes to a more complicated global solution [7]. A simple example is a group effort to approximate a solution of a differential equation using Euler's method. In this structured conversation, each student is assigned a step in the Euler process and needs to choose, on the basis of the results posted by the owner of the previous step, which of the family of parallel line segments matches up to continue the solution. The first segment, of course, has no previous step to match with, providing the opportunity for an initial value discussion. The students in the group can experience the problem's solution arising piece by piece from their individual choices, and can see the effects of a change of initial value rippling through the solution, forcing them to adapt the choices they make. A related, very interesting example which has been used with 4th and 5th graders, but without benefit of the kind of technology envisioned here, is the problem of how the eye perceives shape from a shaded surface [8].

Plugging Into Rich Environments

The Internet (and the Web, in particular), like the real world, has a wonderful cornucopia of rich, immersive, experiences. But, like the real world, many of these experiences can be too rich, too open-ended, with too many pitfalls and dead ends for everyday classroom use where contact time with students is at a premium.

Imagine that to learn about a city you could actually go there, or to learn about jet propulsion you could fly a jet. These imaginings, and more, are becoming possibilities over the Internet. But, does it make pedagogical sense for every student to experience such an immersion for every learning objective? How very useful it would be to be able to actively guide an audience through a website, filtering out the distractions, providing a focus and commentary, providing an experience shaped to the particular needs of each audience. This same concern applies to other data sources, including census statistics, financial and stock market data, data from simulations and web-published papers, etc. While there are a number of experimental systems which attempt this kind of tight integration and "re-purposing" of external content [9], achieving a similar kind of customized delivery of information in mathematics, science and engineering courses remains a challenge.

Assessment

Of all the areas in which technology — both communication technology and computation technology — could have a significant impact, assessment, both formative (feedback and guidance) and summative (examinations, inventories, and rankings) holds the greatest promise and is currently the least well developed. An instructor typically has about six or seven minutes per week per student which can be allocated to assess-

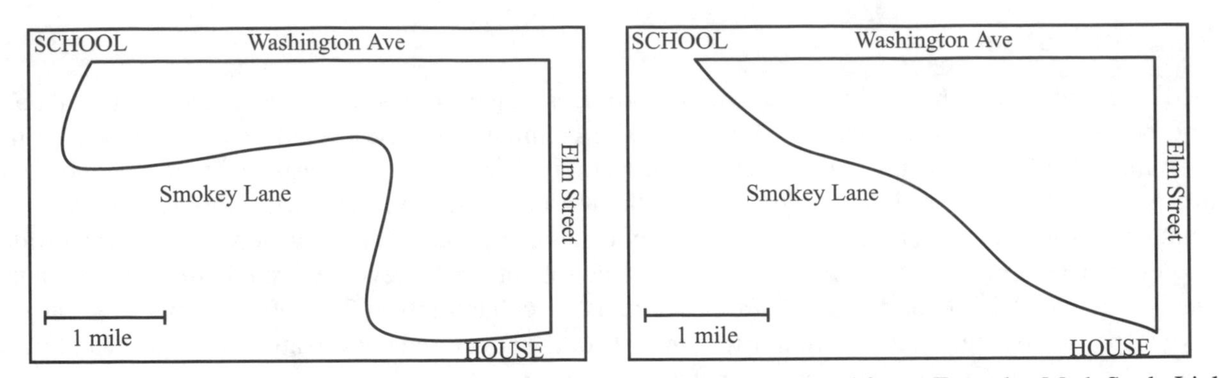

Figure 1. Two distance determination problem instances automatically generated for an Everyday Math Study Link.

ment, feedback, and grading. It's no wonder that multiple choice is such a popular format, and homework and other out-of-class work is generally assigned with reluctance. While some might argue that multiple-choice questions, carefully constructed and used appropriately, can be adequate to the task, no one would argue that they are ideal from a learning perspective. They are, in the current context, a simple necessity.

Perhaps it is the response from technology that needs to adapt in order to facilitate a more efffective assessment environment. It would seem to be a simple matter of programming to build into a calculator the capability to recognize a posed situation such as multiplying two single-digit integers and react with "That's easy. You try it." More generally, technology could be configured to recognize, at least in a rudimentary way, a variety of "learning moments" and react appropriately, effectively extending an instructor's reach, both in space and in time. Moreover, rather than act as a crutch it would enhance student confidence and improve their internal assessment of self-worth.

Writing good problems is difficult. Writing enough variants of good problems to provide a solid practice environment for a diverse population of learners is even more difficult. Allocating enough time to correct these problems and offer effective feedback is nearly impossible. Technology of any sort will not make writing good problems any easier. But if those good problems were designed as templates, technology could be used to produce unlimited numbers of variants of these problems on demand, as well as close the feedback loop on the learner's interaction with those problems.

Everyday Math® Study Link [10] does this rather effectively for various fundamental mathematical topics. Figure 1 illustrates two auto-generated instances of the same problem of determining the distance travelled along Snakey Lane. An important feature of the range of generated versions is that in some versions, Snakey Lane is nearly a straight line and hence, as hypotenuse, shorter than the alternate route, while in other cases, Snakey Lane lives up to its name and is, in fact, longer than the alternative route. The questions posed to the students remain the same throughout all problem instances.

In this web-browser delivered example, the main image is generated from a PostScript® file with curve coefficients generated randomly, but with bounded range. Each time the page is requested, a new problem instance is generated and inserted into the document before it is passed to an image rasterizer (Ghostscript here) to produce an image viewable on a web browser.

The "correct" answers are generated from the same coefficients (exercise: compute the arc length in pixels of a curve segment generated by a ***curveto*** command as a function of its input parameters). When the completed page is submitted, the answers entered are compared with the "correct" answers, with allowances made for the approximation process.

Conclusion

At this point there is, unfortunately, no math engine well suited to all the requirements necessary to allow the kind of automated practice and assessment described earlier to receive widespread distribution and acceptance throughout education. So where does that leave us? Perhaps in closing, it would be helpful to illuminate some of the characteristics needed in a flexible, efficient, and effective mathematics application capable of bringing the promises of technology closer to reality. Nine immediately come to mind:

1. Able to be operated as a server in a request-response manner, handling multiple requests simultaneously.
2. Stateless, so that every request and response can be viewed as independent of every other.
3. Flexible, extensible syntax which on every request can be tailored to the subject, level, and approach.
4. Contains extensible domain-specific mathematical semantics which can reflect the semantics of the subject, level, and approach.
5. Flexible, extensible result formatting which covers the wide variety found in current documents.
6. Able to recognize functional or semantic equivalences (or "near equivalences") in a variety of domains
7. Able to generate random elements from implicitly characterized sets, e.g., 'a random pair of two-digit integers whose product is less than 2000'.
8. Contains indefinite precision numeric evaluation of elementary functions.
9. Contains the usual repertoire of symbolic computations.

References

1. R.Howe, Knowing and Teaching Elementary Mathematics (book review), *Notices of the AMS* **46** (1999), 881–8872.
2. Liping Ma, *Knowing and Teaching Elementary Mathematics: Teacher's Understanding of Fundamental Mathematics in China and the United States,* Lawrence Erlbaum Associates, Inc., 1999.
3. Mathematical Markup Language (MathML™) 1.01 Specification, Patrick Ion and Robert Miner, eds.
4. R.Robson, A Mathematician's Guide to IMS Metatdata,Draft, `http://robby.orst.edu/papers/guide_v2`
5. A nomograph-style rendering of $x \to 1/x$:

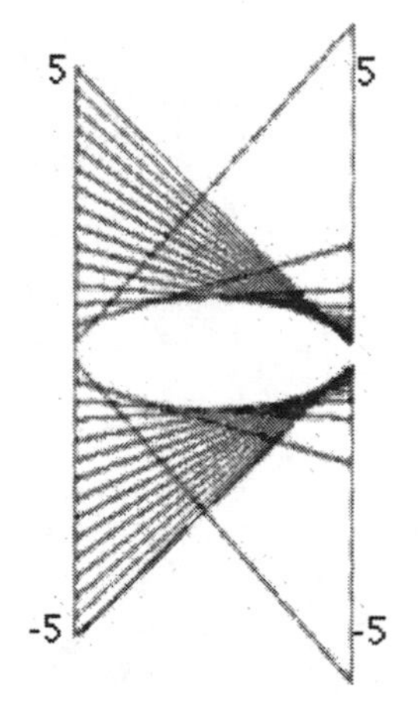

6. `http://www.eoe.org/`
7. Wilensky and Stroup, *Learning through Participatory Simulations: Network-based Design for Systems Learning in Classrooms,* Proceedings of CSCL '99, Computer-Supported Collaborative Learning, Stanford University.
8. Patton, *Shared Dataspace: A Frontier Scout's Report*, Presented at the USMA, November 1999.
9. Ka-Ping Yee, *CritSuite, Critical Discussion Tools for the Web,* Forsight Institute, `http://crit.org/index.html`
10. Max Bell, et al, *Everyday Mathematics, Fifth Grade,* Everyday Learning Corporation, Chicago 1997.

Technology in Education: Minefield or Cornucopia?

L.G. de Pillis
Harvey Mudd College

Abstract. Technology, for the most part, enjoys a positive image among educators. Sometimes, the incorporation of technology in the classroom is simply a result of trends or fashion, but the trend is often justified and fueled by real successes. However, too often the disadvantages are ignored or not well understood. In this paper we elaborate upon the dual nature of technology, its applications, and its uses and misuses in education.

Introduction

There is often confusion between two distinct aspects of technology and its use in education. These different yet related issues are "Technology as an End" and "Technology as a Means". Technology as an End refers to familiarizing students with the fundamentals of technology itself, giving the students the ability to use and master technological tools to achieve their own goals. This could include equipping students with scientific programming skills, with an understanding of the use of high-level mathematics and science software packages, or with the ability to track down useful information through the World Wide Web. At least a superficial understanding of how the tools work and the possible pitfalls one may encounter through the use of these tools is essential to the effective employment of these technologies.

Technology as a Means refers to the use of technology as a tool to achieve our own educational ends as mathematics and science instructors, as opposed to technology as an end in and of itself. In this case, an understanding on the part of the students as to how the technology actually functions is not necessary to the effective use of the particular tool. That is, we are distinguishing between, say, being able to program a Matlab function versus using the results of a preprogrammed function to help elucidate a concept. An instructor could, for example, employ visualization tools to illustrate mathematical concepts, such as real-time rotations of three-dimensional geometric objects, animations which demonstrate the mathematics of change, or plots of solutions of multi-parametric physical models. This category could also include distance learning and other forms of classroom teaching, which incorporate technology. The point is that the focus here is on effectively delivering important concepts in mathematics or science, not on the technology. Such use of technology could serve to enhance a student's understanding of mathematical or scientific concepts without the student ever having to understand how the tool itself works.

We hold that we must be alert to the rapid developments of technology, and should be careful not to view these developments as disjoint from the teaching and learning of mathematics and the sciences. To most, of course, it is axiomatic that as new ideas emerge and develop, they should be woven into the fabric of what we teach. But why is this so? Why can we not keep teaching what we have always taught in the way we have always taught it?

The answer is this: the world we live in is continually changing, and much of that change is due to technological developments and advances. If our goal as teachers of mathematics and science is to prepare our students to function independently and productively once they have left the safe harbor of our educational institutions, we are obliged to equip them with the ability not only to make effective use of technological tools, but with the ability to master new tools and concepts as they are created.

Understanding that it is important to keep students informed of technologies as they emerge, we wish to keep the focus of the rest of this paper on the second sub-category of technology: the use of technology as a teaching tool. In particular, we will focus on the incorporation of computer tools (demonstrations, instructional tools, etc.) into the classroom. This is an area in which some controversy potentially exists concerning the usefulness of technology.

There are some very effective instructors who teach in a completely traditional way and do not employ any technologies at all. Yet they and their students appear to be quite content with the results. The definition of "effective teaching" is somewhat elusive, and whether a student has learned effectively is difficult to quantify. Many agree that the real test of whether learning occurs is not necessarily classroom performance, but whether knowledge and the ability to acquire knowledge is retained well beyond the end of the semester. Since most of us do not have the resources to apply this retention test, we use our instinct and experience to gauge whether what we do within the classroom is beneficial.

We also use the resources we do have at our disposal, for example, class feedback, exams, work-papers, and so forth, to determine whether students are obtaining a deep understanding of fundamental mathematical and scientific concepts during the course of a semester. Moreover, we attempt to gain a clear indication as to whether the student understanding is solid enough that our students will actually *retain* what they have learned.

The goal of employing technological teaching tools is twofold. First, technology aims to make the teaching and learning processes more effective. And secondly, it aims to deliver core scientific concepts to students in such a way that the students do develop a deep understanding and are thereby equipped to acquire and develop subsequent knowledge more independently. There is now an increasing stream of new technological teaching tools being developed and perfected whose goals align with these.

Although there are some who feel there is no call to modify how they have been teaching mathematics or science for the past decade, we hold that the mere existence of this available technology should have some impact on how we teach. We should at least consider and evaluate these new tools as they emerge, determining on an ongoing basis whether a new technological tool could be beneficial to us and to our students in our particular classroom setting.

The Drawbacks and Benefits of Technology

It is not the case that if something is new, it must be good. In the process of evaluating whether a particular type of technology can be useful in a certain classroom setting, we must consider the benefits and possible drawbacks of incorporating these new technologies. Briefly, we see some of the major drawbacks to incorporating new technologies in teaching to be as follows:

- *Time investment*. The learning curve. If the user (the instructor or the student) is not already familiar with the software or hardware to be used, there can be a significant amount of start-up time invested before the user is comfortable with the tools. Moreover, until a comfortable level of understanding of these tools is achieved, the technology can in fact act as a barrier to learning. A well-meaning instructor once tried to introduce the concept of delta-epsilon proofs via the use of a graphing calculator. After expending an inordinate amount of time manipulating commands on the calculator, the instructor came to the realization that one had to first master the delta-epsilon proof in advance of attempting to carry out the related exercises on a calculator. Instead of enlightening, the calculator acted to distract and confuse the central focus of the classroom effort.

- *Ease of use*. Not all classrooms are equally well equipped to handle real-time computational demonstrations or student-computer interaction. Sometimes the practical issues of just getting the equipment set up properly gets in the way of being able to incorporate certain technologies into the classroom setting.
- *Calculation skills lost*. This is an issue that is of serious concern to many instructors. There is the possibility that because our technological tools do so much of the "grunt work" for us, we can too easily forget how to perform some of the simplest tasks ourselves.

Additionally, one should consider issues regarding:

1. the potential rapid obsolescence of a new technology,
2. the possibility of making a commitment to the wrong platform, and
3. the need to have a back-up plan if the technology fails.

Information stored on 8-inch floppy disks is no longer available for retrieval, investment in beta-max video proved worthless, and losing the ability to call up a critical PowerPoint presentation are relevant examples of these three issues.

On the other hand, we feel that many of the benefits of technology in the classroom balance and even outweigh the disadvantages. We view some of the strengths of classroom technology to be as follows:

- *Visualization*. With the aid of computational tools, it is possible to visualize three-dimensional rotations, surfaces generated by two-variable functions, cross-sections of geometric shapes, time-evolution of physical systems, and other graphical solutions to a wide variety of problems.
- *Modeling and Demonstration*. Simulations of physical experiments can be carried out computationally, and systems can be seen to evolve in real-time.
- *Discovery*. Technological tools can allow students to discover scientific and mathematical concepts on their own by removing the need to carry out time-consuming hand calculations, and allowing students to visualize concepts.

Beneficial Tools

Many mathematicians have come to realize the great benefits of using a computer for teaching, both in and out of the classroom. Most of these benefits cannot be achieved in any other way. Mathematicians have become fairly comfortable with the use of packages like Matlab, Maple or Mathematica in the teaching of calculus, performing tasks like generating complicated three-dimensional surfaces which can be rotated in real-time, or summing series on the fly. The ATLAST program (Augment the Teaching of Linear Algebra through the use of Software Tools), headed by Steven J. Leon of the University of Massachusetts, Dartmouth, was created specifically to integrate the use of computation into the teaching of linear algebra. Through this project, a number of effective lesson plans have been developed which introduce or expand on linear algebraic concepts using computer demonstrations.

In an accompanying text, "ATLAST Computer Exercises for Linear Algebra", edited by Steven Leon, Eugene Herman and Richard Faulkenberry, a number of Matlab computer exercises and demonstrations have been collected which also greatly enhance the linear algebra educational experience. One very nice Matlab demonstration visualizes the stretching effect of applying a matrix operator to a circle in the plane and shows graphically where eigenvalues and eigenvectors lie. Another Matlab demonstration generates an animation of a stick-person walking across the screen, using only linear translations and rotations.

In the area of differential equations, Robert Borrelli and Courtney Coleman of Harvey Mudd College were among the first to take a modeling approach to the teaching of this subject in their text *Differential Equations, A Modeling Perspective*, published by Wiley. Most recently, Borrelli, Coleman and others in an NSF-supported consortium (Consortium for Ordinary Differential Equations Education: CODEE) of colleges, universities and industry have created the award-winning package ODE Architect, which is filled with impressive modeling, visualization and solution tools for handling differential equations.

Some of these packages are so easy to use that it seems that a student may simply forget the underlying mathematics upon which the packages were built. To some degree that is possible. However, it is also important to alert students to the fact that although we have a host of powerful computational tools at our disposal, we must not blindly use black-box software without being aware of possible pitfalls. And this means understanding some mathematics. Solving a non-stiff problem in ordinary differential equations using a stiff solver inherently sacrifices accuracy. On the other hand, applying a non-stiff solver to a stiff problem introduces the risk that the algorithm may never converge to a solution!

Outside the classroom, students use computers to solve problems that reveal instructive concepts in the broad structure, but which would become overwhelmingly tedious and non-enlightening if the student had to carry out hand computations. Some examples include the solution of least squares problems, the threading of cubic splines through discrete data points, the optimization of constrained problems, or the solution of systems of ordinary differential equations. Computers are also used for independent discovery. One can develop computer activities that lead students to evolve their own theories based on observed trends. As an example of exercises which guide the student through such a discovery process, we note the work of David Hill and David Zitarelli of Temple University who have developed a text containing appropriate computational labs for discovering linear algebraic concepts. The labs are collected in the Prentice Hall publication, "Linear Algebra Labs with Matlab".

These are but a few of many examples of the rich array of uses instructors have found for enhancing the educational experience of mathematics students through the use of technology.

Conclusion

As technology becomes more accessible and user-friendly, our concerns about start-up time investment and ease of use of technology become less important. Even now, there are classrooms in which students are expected to bring their own graphing calculators or even laptop computers. Little or no time need be spent in setting up error-prone projection systems. This can allow the instructor to guide the students through interesting computational demonstrations with very little start-up time. This can also allow for students to modify demonstrations to answer their own questions. More student interaction in the classroom is thereby encouraged, since the students have access to their own set of computational tools.

The issue we should be most conscious of in the context of increasing use of technology in the classroom is that of loss of basic computational skills. Even now, for example, it is not unheard of that students who complete certain types of calculus classes do not have the skills to employ even the most basic of computational techniques for integrating special functions. However, as long as we remain aware of this possible pitfall, and take steps to counteract it, we should be able to safely make increasing use of technology in the classroom to the overall benefit of both instructors and students. A thorough understanding of the Minefield of Technology is essential to being able to reap the benefits of the Cornucopia of Technology.

The Impact of Technology on Experimentation and Its Implications for How and What Students Learn in Calculus

Frank Wattenberg
United States Military Academy

Abstract. Modern technology provides students and educators with a rare opportunity to bring experimentation to life in a classroom environment. We present several illustrative examples of using experimentation to complement student learning of associated mathematical topics using resources available through commercial means and freely accessible on the World Wide Web. Additionally, we present a general philosophy focusing on how one might effectively and efficiently integrate experimentation into their curriculum.

Introduction

Modern technology—including handheld technology, computers, and thc World Wide Web—has dramatically expanded the kind and amount of experimentation that is feasible in the classroom. Virtually every student now has at his or her fingertips sophisticated and compelling simulations, inexpensive, flexible equipment for "bench-lab" or "wet-lab" experiments, remotely accessible experimental equipment, massive and real-time data sets; research- and museum-quality primary resources; and new kinds of Web-mediated distributed experimentation.

This paper focuses on experimentation in the various modes listed above in the context of widely available and powerful computer-based tools, including Java applets, computer algebra systems, and modeling packages. Throughout we stress the importance of flexible, general-purpose tools that enable students to explore ideas of their own beyond the confines of pre-programmed applets so that students are better prepared to do the open-ended experimentation and analysis required in the workplace. We also discuss the implications of such an approach for how students might best learn the subject of calculus and what topics are most important both for a first year, possibly terminal, calculus course and for a two-year core mathematics curriculum. The two most important points we wish to make are related – the importance of modeling and the importance of experimentation and data analysis to facilitate student understanding.

We begin by introducing two examples that represent the essence of this pedagogy, and then proceed to draw some lessons learned by these experiments. Finally, we close with some comments on the implications these technologies hold for the future of mathematics education.

Experimentation

For our first example, consider two separate modeling situations. The illustration in Figure 1 shows two cylindrical tanks that are open at the top and connected at their base by a tube so that the contents of the

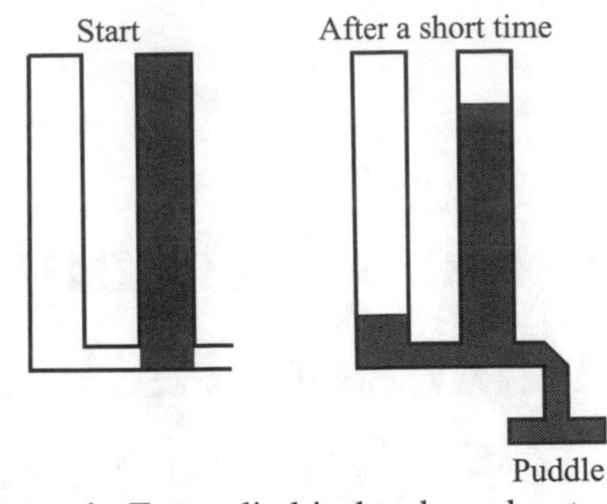

Figure 1. Two cylindrical tanks and water

two tanks can flow freely between them. The right-hand tank also has a tube or drain at its base that allows its contents to spill onto the floor. We begin this experiment by filling the right-hand tank with water. After a short time, some of this water has spilled onto the floor and some has flowed into the left-hand tank. We ask the question — Will the water level in the left-hand tank ever be higher than the water level in the right-hand tank?

Figures 2 and 3 illustrate a circuit with two capacitors and two resistors. To initiate this experiment, we apply a charge to the right-hand capacitor. After a short time some of the charge from the right-hand capacitor has leaked into the ground and some has leaked into the left-hand capacitor. We ask the question—Will the charge on the left-hand capacitor ever be higher than the charge on the right-hand capacitor?

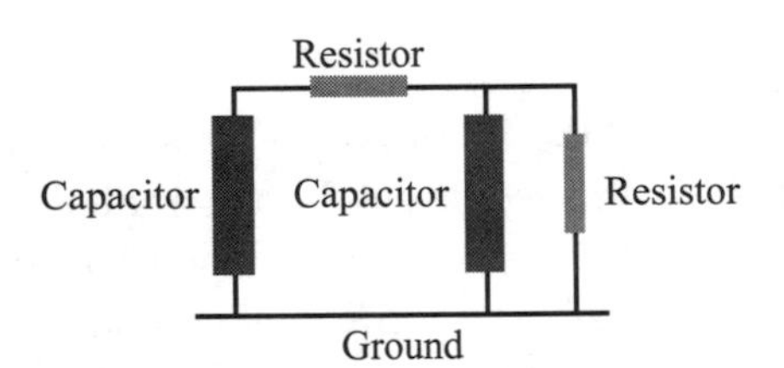

Figure 2. A schematic of an experiment with capacitors and resistors

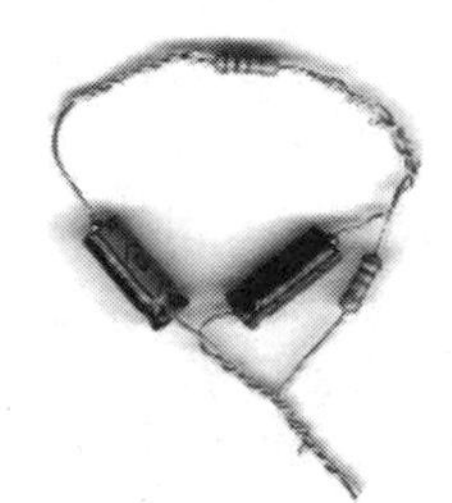

Figure 3. An experiment with capacitors and resistors

We might model the resistor and capacitor situation by the system of differential equations

$$\frac{dL}{dt} = a(R - L)$$

$$\frac{dR}{dt} = b(L - R) - cR$$

where R is the charge on the right-hand capacitor and L is the charge on the left-hand capacitor. The values of the constants a, b, and c are all positive and depend on the physical characteristics of the components of the experiment—capacitance of the two capacitors and the resistance of the two resistors (Question—Does the same system of equations model the first situation?) This particular experiment has intrinsic appeal because the mathematics involved in this system of equations is accessible in a first-year course: the meaning of the equations; numerical approximations to solutions; exact solutions; and qualitative analysis. In addition, inexpensive and versatile equipment like the Texas Instruments Calculator-Based Laboratory put these experiments within easy reach.

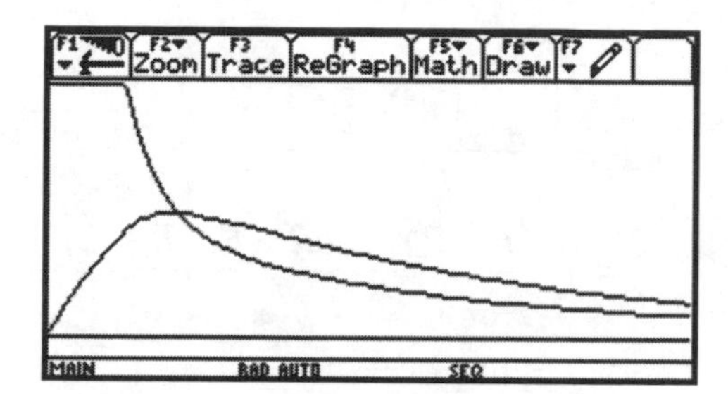

Figure 4. The results of a physical experiment

Figure 4 shows the results of one experiment using an actual capacitor and resistor circuit in the configuration described earlier. The right-hand capacitor was charged by touching a wire from a battery to the capacitor's lead. The touch was not instantaneous—hence, the flat portion of the curve. Notice the charge on the right-hand capacitor starts out high and drops steadily toward zero. The charge on the left-hand capacitor starts at zero, rises until it is above the charge on the right-hand capacitor and then begins to drop. Figure 5 shows a numerical approximation to one particular example of our system of differential equations.

The exact solution of this system of differential equations is also easily accessible in a first year calculus course. Rewriting the system of equations as

$$\frac{dL}{dt} = -aL + aR$$
$$\frac{dR}{dt} = bL - (b+c)R$$

we know that the solutions are of the form

$$L = c_{11}e^{\lambda_1 t} + c_{12}e^{\lambda_2 t}$$
$$R = c_{21}e^{\lambda_1 t} + c_{22}e^{\lambda_2 t}$$

where 1and 2 are eigenvalues of the matrix

$$\begin{bmatrix} -a & a \\ b & -(b+c) \end{bmatrix}.$$

We can obtain additional insight into this system of differential equations by looking at the variable

$$v = \frac{R}{L}.$$

Again assuming that all variables are functions of t, we can differentiate v so that

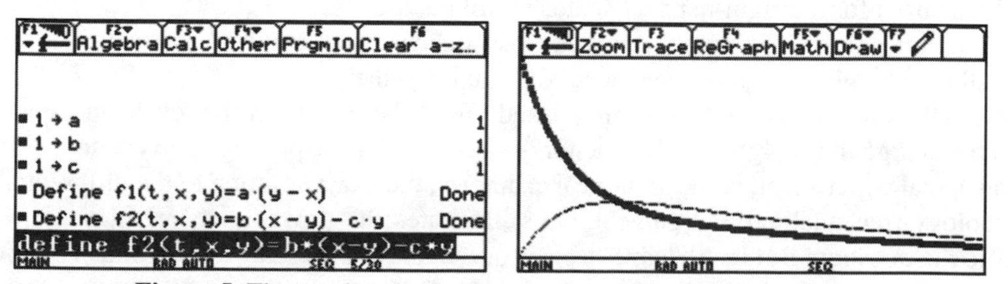

Figure 5. The results of a numerical approximation to one solution

$$
\begin{aligned}
v' &= \frac{L\left(\frac{dR}{dt}\right) - R\left(\frac{dL}{dt}\right)}{L^2} \\
&= \frac{Lb(L-R) - LcR - Ra(R-L)}{L^2} \\
&= b - bv - cv - av^2 + av
\end{aligned}
$$

This is an interesting exercise in applying the Quotient Rule. This now allows us to examine the differential equation or one-dimensional dynamical system

$$
\frac{dv}{dt} = p(v)
$$

$$
p(v) = -av^2 + (a-b-c)v + b
$$

The graph of $p(v)$ is U-shaped with the U-opening downward. Notice that $p(0) = b$ is positive. Thus, $p(v)$ has one positive zero, v^*; is positive for positive values of v to the left of v^*; and negative to the right of v^*. Since v is never negative, v^* is an attracting equilibrium point. Next, notice that $p(1) = -c$ is negative. Thus, v^* is less than 1. This implies that as t approaches infinity v, or R/L, approaches a limit that is less than 1. In other words, the charge on the left-hand capacitor will always eventually rise above the charge on the right-hand capacitor.

We can do better yet by using a technology-free combination of physical intuition and qualitative analysis of the original system of differential equations. Figure 6 shows the beginning of a rough sketch showing the charge on each of the two capacitors. In the beginning the charge on the right-hand capacitor falls as it drains into the ground and into the left-hand capacitor The charge on the left-hand capacitor rises.

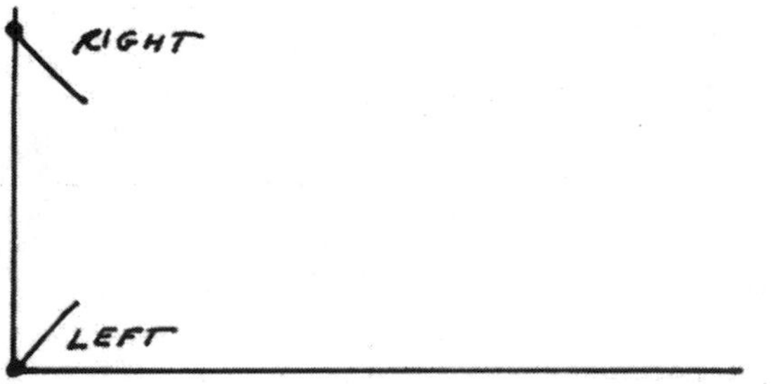

Figure 6. The beginning of a rough sketch

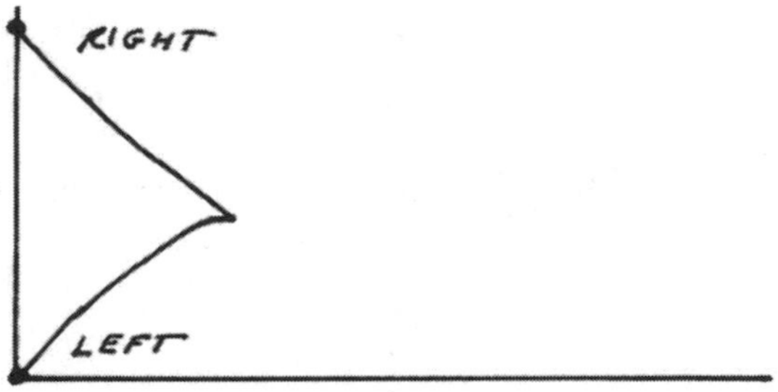

Figure 7. The rough sketch continued

Figure 7 shows what happens next. The charge on the right-hand capacitor continues to drop and, as long as the charge on the left-hand capacitor is below the charge on the right-hand capacitor, the charge on the left-hand capacitor continues to rise. When the charges on the two capacitors are equal, the charge on the left-hand capacitor is not changing, but the charge on the right-hand capacitor continues to drop as it drains onto the ground. That is, when the two curves cross, the curve for the left-hand capacitor is horizontal and the curve for the right-hand capacitor is still decreasing.

Figure 8 continues the story. After the curves cross, the curve for the left-hand capacitor will decrease. Eventually all the charge from both capacitors will drain into the ground.

Mathematically, the answer to the question asked—Will the charge in the left-hand capacitor ever be higher than the charge in the right-hand capacitor?—is somewhat surprising and very satisfying. It is supported by an actual experiment; some numerical examples; the exact solution to the differential equations; and a technology-free qualitative argument. This illustrates the power and interplay of four different approaches to the question. But it still leaves a basic question unanswered—How good is our model? Does it apply to the water situation as well as the electrical one? If so, under what assumptions? For example,

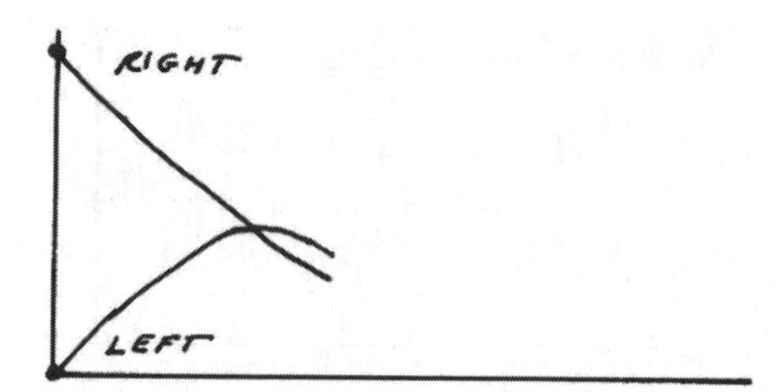

Figure 8: The rough sketch continued some more

suppose that the tube connecting the two tanks is very wide and the tube draining the right-hand tank onto the floor is very thin. Then the water might flow back-and-forth between the two tanks and this kind of model might not be appropriate because it fails to take into account the momentum of the water.

A Second Experiment

This example is built around three "bench-lab" experiments using inexpensive readily available equipment, one simulation-based experiment, and investigations using a computer algebra system—all of which study interference and diffraction.

The first experiment uses a laser pointer and finely ruled slides to produce diffraction patterns. Figure 9 shows a crude diffraction pattern that was produced using "slides" made on an ordinary Postscript laser printer.[9]

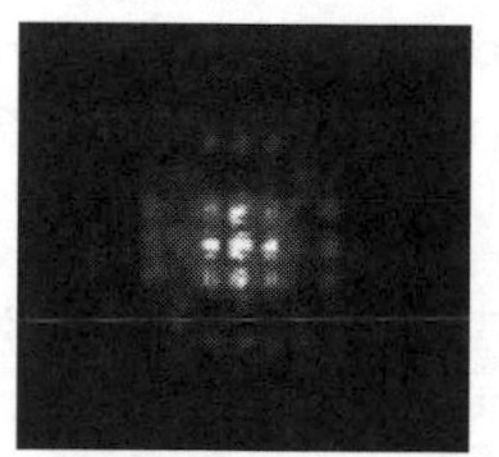

Figure 9. A crude diffraction pattern

The second bench-lab experiment uses two battery-powered speakers with a portable cassette player or a notebook computer to generate a 440 Hz tone. If the two speakers are placed outside about three feet apart, walking along a line as shown in Figure 10 will enable one to hear the interference pattern produced.

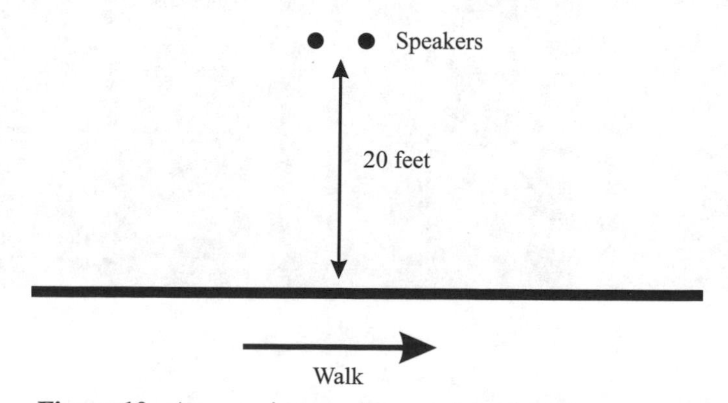

Figure 10. An experiment with sound and interference

[9] These slides are downloadable from http://umastr1.math.umass.edu/~frankw/ccp/GraphPaper/diffraction/index.htm An even better slide, part of an excellent unit on the structure of DNA, is available by mail from the Institute for Chemical Education at htttp://jchemed.chem.wisc.edu/ice/

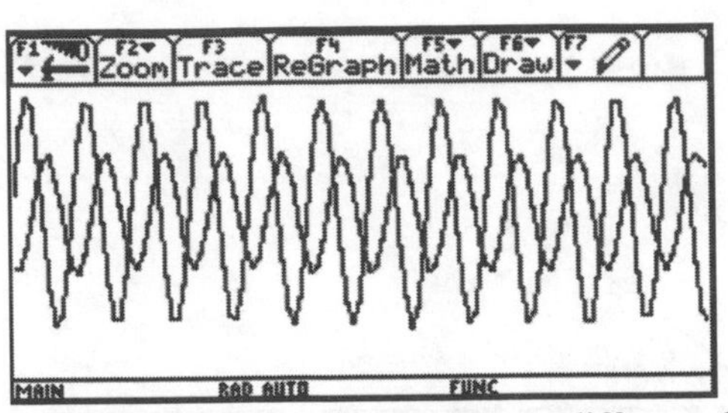

Figure 11. The sound recorded by two microphones at different distances from a source

At the point on the path that is closest to the speakers, the two speaker signals will reinforce each other producing constructive interference. Eventually, one will reach a point where the two signals are 180 degrees out of phase and *almost* cancel each other out—destructive interference. Continuing past this point will cause the signals to again interfere constructively.

The third bench-lab experiment uses a Texas Instruments CBL and two microphones to examine the sound signal received by two microphones at different distances from the same source. Figure 11 shows one typical result. Notice that the signal recorded by the microphone that is further away lags behind the signal recorded by the closer microphone. This distant microphone also has lower amplitude, as expected.

The simulation-based experiment employs a virtual ripple tank. Two such Java applets are shown in Figures 12 and 13[10].

We can use each of these tools in various mathematics courses in different ways. One underlying message is the importance of mathematics. Mathematics is key to our understanding of lasers, acoustics, diffusion, and interference and it played a key role in the discovery of the structure of DNA. In addition to that subliminal point, we want to make two important points here—the importance of understanding modeling and the importance of general-purpose modeling tools, including computer-algebra systems.

Although the ripple tank applets are very impressive, they are only approximations. Neither simulation takes into account the effects of distance on amplitude, a component of the physical reality that was readily apparent in the bench-lab experiment. The simulations predict complete cancellation or 100% destructive interference. In experiments with light and/or very finely ruled slides, however, these simulations are very effective models because the distances are so large compared to the wavelength of light (physicists call this far-field interference). Our first experiment with sound, however, cannot be modeled effectively without considering the effects of distance (near-field interference). After students look at the clean pre-

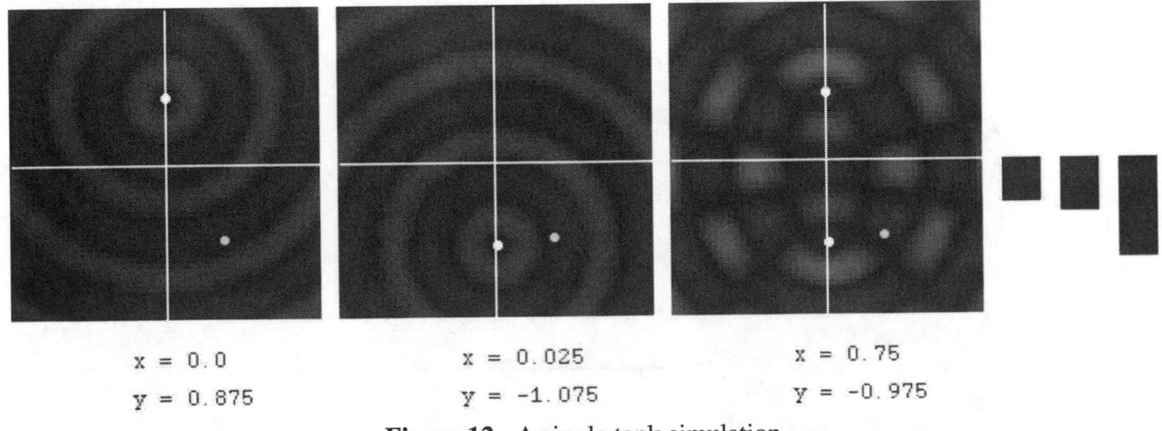

Figure 12. A ripple tank simulation

[10] This JAVA applet is available from the WebPhysics Project at http://webphysics.davidson.edu/Applets/Applets.html

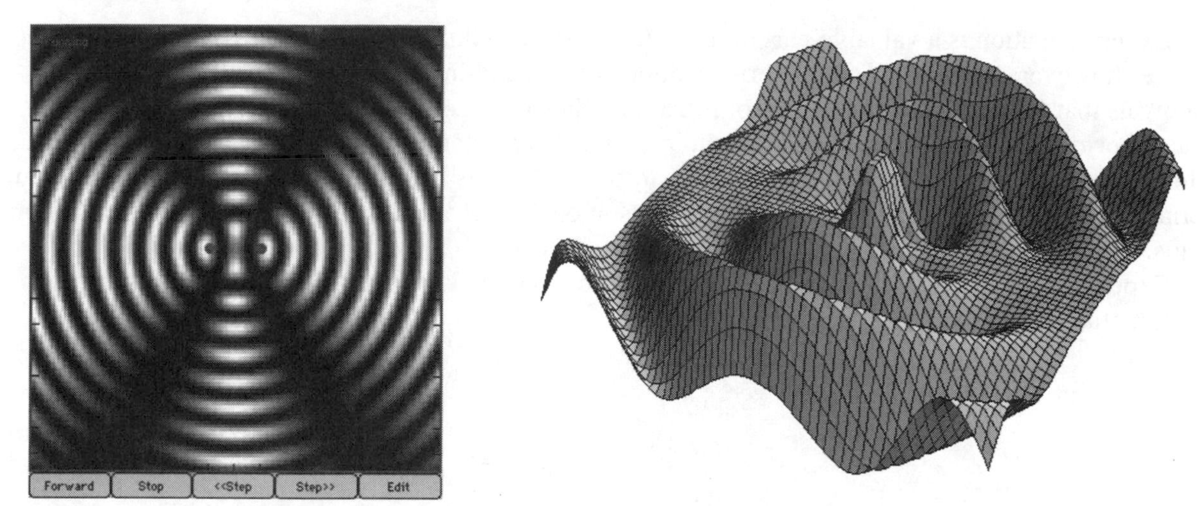

Figure 13. A second ripple tank simulation

Figure 14. One frame of a CAS animation

dictions of a ripple tank applet and then experience the less than 100% (but still impressive) destructive interference of the first experiment with sound, they gain an appreciation for some of the subtleties of modeling. Better yet, they can use a computer algebra system to produce their own models that take into account the effects of distance.

Figure 14 is one frame from an animation produced by a computer algebra system. This CAS animation is visually more impressive than either of the ripple tank applets and students can modify it to add the effects of distance.

There are some important trade-offs between dedicated simulations like the ripple tank applets and more general-purpose tools like computer algebra systems. With the former, the technology is often flashier and less intrusive. Students stay "on task" and see what we want them to see. With the latter, students must do more work and they may be distracted by the technology, but they have the flexibility to answer questions of their own and they are learning general-purpose skills.

Implications

The environment in which our students will live, work, and do mathematics has dramatically changed. Our students have newfound access to a startling array of experiments and experimental results. In some cases, they have access to research quality equipment, simulations, and primary data from respected sources. In other cases, the provenance is less certain and the documentation is incomplete. Even research quality data is often subject to different interpretations. In addition, students have access to professional quality tools for analysis, visualization, and model building. This cornucopia of possibilities will continue to grow as students move into their careers and as the power of technology continues its explosion.

> Modeling, probability and statistics, differential equations, and multivariable calculus are so important that along with calculus they are an essential part of the college education of every student. Computer-based skills are enormously important.

The most important single factor affecting the quality of college-level mathematics learning is the time that students spend on learning and using mathematics. This implies both that more time should be spent in mathematics classes and that classes in other subjects should routinely use mathematics. In this setting, modeling and applications are essential complements to student understanding. In this setting, modeling and applications are essential complements to student understanding. Realistic applications with their essential complexity impose standards that are more natural, more acceptable to students, and higher than those imposed by more traditional textbook problems.

Experimentation is a valuable part of mathematics courses and learning mathematics. Experimentation alone, however, is insufficient. Experimentation works hand-in-hand with theory. Experimentation can show us that for certain combinations of resistors and capacitors in Example 1 the charge on the left-hand capacitor does rise above the charge on the right-hand capacitor but it cannot tell us, as theory can, that this always happens. But – wait – theory can only tell us that this always happens when our model is appropriate. Experimentation can help us test the limits of our model and develop new models when this model fails. Either experimentation or theory alone is insufficient.

Experimentation should always be guided by theory and by knowledge of the area of application. The differential equations

$$\frac{dL}{dt} = a(R - L)$$
$$\frac{dR}{dt} = b(L - R) - cR$$

that we used in Example 1 came from a reasoned hypothesis about the behavior of the underlying physical situation. The success of this model gives us confidence in this hypothesis and increases our physical understanding.

The ease and power of technology can make a wild trial-and-error approach to modeling or an approach based on unreasoned data-fitting seem attractive. But these approaches are, at best, only useful as first steps that should be followed by a deeper understanding of the models suggested by trail-and-error or data-fitting. For example, students might observe by data-fitting that pressure and temperature are related by the equation

$$P = kT,$$

where k is a positive constant and T is temperature in degrees Kelvin. Observing this relationship by itself, however, does not imply that students understand that increasing the speed of gas molecules increases both the number of times per second that a gas molecule hits a surface and the momentum transfer that occurs with each such collision. The result is that pressure is proportional to speed squared. Understanding the relationship above follows from the further understanding that temperature in degrees Kelvin is a measure of the kinetic energy of the gas molecules which, in turn, is proportional to speed squared. Thus, the real power of technology comes not from technology by itself but rather from its use together with more traditional mathematics.

Conclusion

The impact of both current and future technology depends more on how we use it than on the technology itself. As educators, our first concern should be that all our students develop the ability to use mathematics. We must insist that the quality of learning and its accessibility are critical driving forces as we make choices about technology. We should invest in core capabilities—hardware that is capable but not extravagant and software and operating systems that are reliable and do not squander resources.

Perhaps, most importantly, the real impact of technology on our world will depend on whether we use technology to extend our ability to work collaboratively and to make informed, thoughtful, and ethical choices as individuals and as a society. The World Wide Web, together with inexpensive, portable, and flexible laboratory equipment like the Texas Instruments CBL, offers some exciting new possibilities that extend the existing concept of classroom even further. The Web enables educators to conduct large-scale distributed experiments in which a multitude of geographically and temporally distinct sites record individual experimental trial data. Compiling this information then produces a rich source of data possessing a quality of variation far superior to that capable of being assembled by a single class or student group. Additionally, by making this data compilation available to all participants regardless of their percent contribution to the whole, the collaboration compensates for inequitable resources at the individual sites. And this possibility alone provides enough potential payoff to make an investment in effort worthwhile.

Computing, Information, and Communication Technologies: Impacts on Undergraduate Mathematics

John L. Scharf
Carroll College

Abstract. When students have an opportunity to apply mathematics to problems that are real, relevant, and of personal interest, they become self-motivated learners eager to demonstrate their understanding of mathematical topics. We describe the central philosophy concerning how our curriculum redesign efforts have transformed our program so that it welcomes the diversity of experiences that various academic majors represent. By effectively incorporating modern computing technology into our mathematics curriculum and by letting interdisciplinary problems motivate the underlying mathematics, students have risen to new heights of achievement. We conclude that the resulting program has shifted the locus of challenge for the student from being centered in difficult calculations to being centered in modeling, communication, and other intellectually vital activities.

Introduction

For Andrea, Jennifer, and Peter there is no doubt about the value of their mathematics education. Presenting the results of their investigation into bungee cord jumping, they feel good about what they have accomplished. All three are first year students. Andrea is a computer science major, Jennifer is a double major in chemistry and mathematics, and Peter is a civil engineering major. After it is all over, the presentation complete, Peter comments that he has never had so much fun learning. Jennifer and Andrea ask when they will have another opportunity to "show their stuff." The presentation is for a group of campus visitors, high school students and their teachers from across Montana.

Andrea, Jennifer, and Peter are using a spreadsheet with imported mathematical formulations, solutions, tables, and graphics. The look is professional and the results are impressive. They explain to the audience how they performed an experiment on a bungee cord to obtain data that they used to formulate their differential equation model for the bungee cord jump. They detail how they solved the DE and generated graphs of the solutions including phase diagrams using *Mathematica*, and how they tested their results. The challenge they were presented with was to determine how long a piece of bungee cord from a local hardware store would have to be cut so that a 200 gram mass would just "kiss" the floor when dropped from an arbitrarily specified height. This scenario was intended to mimic a contemporary television commercial in which a bungee cord jumper dips a corn chip, held in his mouth, into a bowl of salsa, located on the ground, without bumping his nose.

The campus visitors challenge Andrea, Jennifer, and Peter with a fall distance of 1.6 meters. They do some quick calculations, cut the cord to length, and tie on the weight. The weight is dropped and falls to within one-half of a centimeter from the floor. The visitors are impressed not only by the good results and professionalism of the presentation, but also by the depth of the students' understanding. One of the visiting teachers comments that he did not encounter such sophisticated problems until he was in graduate school.

Technologies Make It Possible

Students can address more sophisticated problems earlier because computing, information, and communications technologies make them accessible to undergraduates. Bungee cord jumping, design of an automobile suspension, lake pollution, the bison in Yellowstone Park, the spread of AIDS, computer-assisted tomography, RADAR detection, the global positioning system, and many other problems can now be tackled by undergraduates commencing in their first-year mathematics courses. High speed, desktop and notebook computers and hand-held calculators are ubiquitous. Spreadsheets, numerical packages, computer algebra systems, as well as simulation and modeling software provide extensive resources for mathematics students. The Internet provides students unprecedented access to information including databases and datasets. And finally, students are communicating with the aid of visualization and presentation software, and they can collaborate with others worldwide about their work via the Internet.

Promises for the Future

In the old days (i.e., five or more years ago) we taught and learned undergraduate mathematics with promises for the future. "You need to know calculus and differential equations so that you will be able to learn physics, engineering, economics and/or more advanced mathematics." The problems were simplistic and contrived—they had to be. For many students the promises were empty, especially for those enrolled in our service courses. They saw through the simplicities and motivation was lacking. The best example of this is the course that has come to be known as pre-calculus.

Up until two years ago, the vast majority of students in pre-calculus at Carroll College were business majors who were fulfilling a general education requirement. This was typically their terminal course in mathematics. While pre-calculus was taught as if to prepare students for subsequent studies in calculus, in reality, it was pre-nothing. Last year we changed our course offerings to add a course in discrete dynamical systems, which includes the study of systems of difference equations using matrix algebra. Now many of the students who would have taken pre-calculus take the discrete dynamical systems course instead. The problems are real and relevant, with many (but not all) of the applications taken from business and economics. This change was possible because computing technology and spreadsheet software enable students to address, with relative ease, realistic problems related to their major interest. It seems ironic that in many ways our new discrete dynamical systems course is better preparation for those students who wish to take calculus than our pre-calculus course is. It is our hope that many students who would normally not take calculus will be motivated to do so by their experiences in discrete dynamical systems.

Today, students gain an understanding and appreciation for the importance of mathematics as they learn it because they *experience* it. Technology allows them to engage real, contemporary, and relevant problems. Mathematics still holds promises for the future, but now it has relevancy in the present as well. The study of mathematics motivates students to learn, it changes their worldview, and it enables them to pursue with confidence a wide variety of interests using newly acquired skills and knowledge.

Communication and Access to Information

Students in elementary statistics now access databases and obtain data sets from a myriad of Internet sites including the World Health Organization, the Center for Disease Control, and the U. S. Census Bureau. They download data sets and use a statistical package and/or a spreadsheet to perform statistical analyses of the data. From this they draw interesting and relevant conclusions and present their work, often using electronic media such as Power Point or a website of their own. Projects have included studies of poverty in America and the spread of AIDS worldwide, among others.

Some of our faculty members now require students to develop webpages as part of a project, and these are posted on a class website. (I know from personal experience that students are much more responsive to developing a webpage than they are to writing a paper or a report.) Many parents are quick to respond and often send us email notes expressing pride in the work done by their children. In addition to my own webpage, other sites to check for periodic postings of student works are Terry Mullen's and Marie Vanisko's webpages at `www.carroll.edu`.

When studying numerical integration in calculus, I have students access current, up-to-the-minute stream flows that are collected at stream and river monitoring stations from around Montana and transmitted via satellites to the Montana Natural Resource Information System website at

`http://montana.usgs.gov/www/rt/imagemap/rt_imagemap.html`.

The data is provided graphically as well as in tables. Students are asked to estimate the net accumulation of water due to stream flows on the Missouri River in a string of three reservoirs near Helena over a specified time period. During this project students are often found checking on streams near their hometowns.

Different but Not Any Easier

Computing technology has made the study of mathematics different, but not any easier. There are new challenges for all of us who study mathematics. Once we learn to effectively use computing software like *Mathematica*, for example, (which in itself is no small task) we can readily obtain solutions for more substantial problems. But, how do we know whether these solutions are correct? How do we test them for validity?

Today's students may not spend as much time learning methods of integration as we did in "the old days", but now they need to learn how to use computers effectively and test their results. In essence, however, the challenges are the same. After all, how does one know that an antiderivative calculated by hand is correct? (You check the answer in the back of the book, of course.)

Today's students are adept at using computers and accessing information on the Internet. They have grown up with these technologies and are generally uninhibited in using them. In fact, most seem to enjoy it. The challenge is to help them learn to use these technologies effectively.

Position on Technology

In 1993, we initiated an effort at Carroll to revitalize our mathematics curriculum. We set the following goals:

1. To integrate mathematical topics so that students would see and appreciate the connections and unifying themes among seemingly disparate areas of mathematics.
2. To motivate mathematics with applications drawn from a wide variety of disciplines in the sciences, engineering, computer science, the social sciences, and the humanities, so that students would see the study of mathematics as relevant and thereby change their worldviews.
3. To integrate the use of technology throughout the mathematics curriculum so that students could access significant and important problems and use mathematics effectively in trying to solve them.
4. To present perspectives from other disciplines in mathematics courses so that students would see the relevance and usefulness of mathematics and the connections mathematics has to most if not all areas that are of interest to them.
5. To include problems for students that require them to address ethical, social, and aesthetic issues in their study of mathematics.

It is this set of goals that embody my position about the future of mathematics education and the role that technology should play.

Over the past seven years, we have had considerable success in achieving our goals and we are proud of what we have accomplished. It is our most sincere hope that our students are the beneficiaries. I often ask beginning students whether they like mathematics. Almost always, the answer is yes. I sincerely believe that most students come to us liking mathematics and desiring to learn more. It is our responsibility to nurture these likes and desires, building more confident problem solvers and better thinkers. Computing, information, and communication technologies make this job much easier.

In my twenty-three years of teaching college mathematics, I have never seen a better spirit and enthusiasm among our students than there is now. Andrea, Jennifer, and Peter are now juniors, progressing toward the realization of their personal goals, and mathematics is an integral part of what they have learned and accomplished. I am confident that their mathematics education will serve them well after they leave Carroll to pursue their professions. Their worldviews will be changed and mathematics will be an integral part of how they think and how they approach and solve real-world problems. The use of technology has been a primary catalyst in making this possible.

Technology and Curriculum Structuring

Patrick J. Driscoll
United States Military Academy

Abstract. We present a general framework of how educational technologies can have an impact upon mathematics curriculum development, illustrating various relationships that exist between student and faculty perspectives. Mathematics education exists within an environment of changing students and faculty, changing technology, and changing expectations. Rather than constraining curriculum design by iterating from traditional textbook content, we propose that educators must now think of courses and curriculum in a novel way that uses technology to maximize the curriculum's ability to deliver customized, relevant content that meets the needs of students now and in the future.

Introduction

Technology has once again jarred the walls of education, continuing a trend begun many years ago with the introduction of pencils, mass-produced paper, the abacus, electricity, television, cameras, and handheld calculators. As in the past, the proliferation of technology throughout our society has a direct impact on the worldview held by our young simply by shaping the environment in which they live. High school students participate in financial investment clubs that, through easy access to computing and communications technology, actively trade securities in real time. They literally have more computing power in their graphics calculators, regardless of manufacturer, than was built into the control systems of any of the Apollo spacecraft. These same students are inundated daily with news of mega-millionaires no more than ten years their senior who pursued entrepreneurial endeavors rather than "cradle-to-grave" professional careers. It is rare today to find a single high school student who does not possess a pager, a cellular telephone or a wireless PalmPilot device; and who expect connectivity and access to information now not later.

While on one hand we can admire the zealousness at which these young minds embrace technology as "business-as-usual," we must also recognize that these are the same people arriving on the doorstep of our educational institutions each year. And, they are doing so conditioned with a high level of expectation concerning the efficiency and potential promises of technology based largely on their limited experience to-date. Since the process of learning is inherently inefficient, educators are placed in a precarious predicament characterized by questions of the following ilk. "Should we de-emphasize or discard completely the comparably arcane student tasks and activities that have traditionally managed to facilitate student understanding of mathematical topics, replacing them with technological efficiencies?" "Or, does the demanding experience associated with, say delta-epsilon proofs, encourage mental acuity or motivate conceptual insights that technology cannot?" "If there is a balance to be struck between these two seeming extremes, where is it?" "Does this balance point differ with each mathematics topic, or is there a single framework that applies across a complete curriculum?"

The last question is one central to this paper. As computing technology saturates our educational environment, pursuing a balance between incorporating efficiencies and allowing conceptual ideas to properly ferment and percolate in each student's mind forces us to constantly examine our mathematics curriculum, our teaching methods, and the very underpinnings of our teaching philosophy. If, as many have put forward, our true service to society is to produce competent, confident problem solvers within this quagmire of technological advancements, then this task must be made central to our means. We present in this paper a collection of thoughts on the resulting impact on the way we do business with the intent of stimulating ideas on where one might look to formulate global curriculum strategies for successfully operating within this environment in the future.

Simply for convenience, in this paper we principally refer to technology associated with electronic hardware devices and computer software applications, since these are the technologies most closely wed to mathematics curriculum reform. However, in our current era, we note that "technology" easily includes items such as wind tunnels, vehicles, robotics, aircraft, transportation and delivery systems, bio-mechanical devices, lasers, DNA sequencers, and the like, which evoke hands-on experimentation more so than aforementioned technologies ever will.

What Experience Has Shown

Reflecting upon the various experiments with technology we have attempted over the past ten years or so, there appear to be several valid generalizations that are possible to make. Since our goal is to shape a broadly defined abstraction of the role of technology in curriculum, it is worthwhile to use these generalizations as the basis for abstraction. With this motivation in mind, let us proceed.

The incorporation of technology into our pedagogy is not, has never been, and will never be, *efficient* when one compares the amount of start-up effort expended by a faculty member to the results achieved by students when viewed using traditional evaluation instruments (written tests, etc.). This does not imply that technology is not making a significant contribution to student learning; it merely suggests that the intellectual progress being made is not observed using these instruments. Technology relieves students of time-consuming, mundane tasks such as repetitive calculation and symbolic manipulation, and shifts the learning paradigm to one of *process*, not product per se. Experimentation, methodology, organization of effort, modeling, refinement, and communication are salient in such a process, as opposed to calculation, memorization, and repetition. Perhaps, as John Scharf suggests in his article that appears elsewhere in this volume, assessment instruments must change if technology is going to be an essential component of learning. The self-motivating and enthusiastic response he observed from students when they are assessed using presentations, interaction with clients, and hands-on experimentation for validation of model results provides us with cogent evidence that such process assessment instruments exist.

The use of technology will always be considered as adjunct to course content, a "fifth wheel" so to speak, so long as its use is orthogonal to the main effort of a course as perceived by students and faculty. When conceptual development of ideas must temporarily halt whilst students shift their focus to tangentially related machine tasks or software features (e.g., demonstrations), technology is not integrated into the course. True integration of technology encourages students to reach for it naturally to continue the pursuit of ideas, not to offer a distraction. The challenge to educators is to help students understand when technology utilization is appropriate and when it is not.

Today's students appear to be clever enough to either learn technology on their own, or find seemingly efficient, perhaps unorthodox, means of minimizing their use of technology to the point of achieving an academic performance level they perceive is acceptable for themselves, whether this be high or low. This may seem an odd generalization given the presumption that students attend college to learn as much as they can in order to advance their station in life. However, in as much as students also perceive college as a vital component in the development of their socialization skills, experience has shown that they develop unique task prioritizations within their individual time management systems. They are extremely reluctant to alter this prioritization so long as it does not catastrophically fail them. The upshot of this observation is that technologies possessing significant learning curves are shunned in favor of those that do not. Hence,

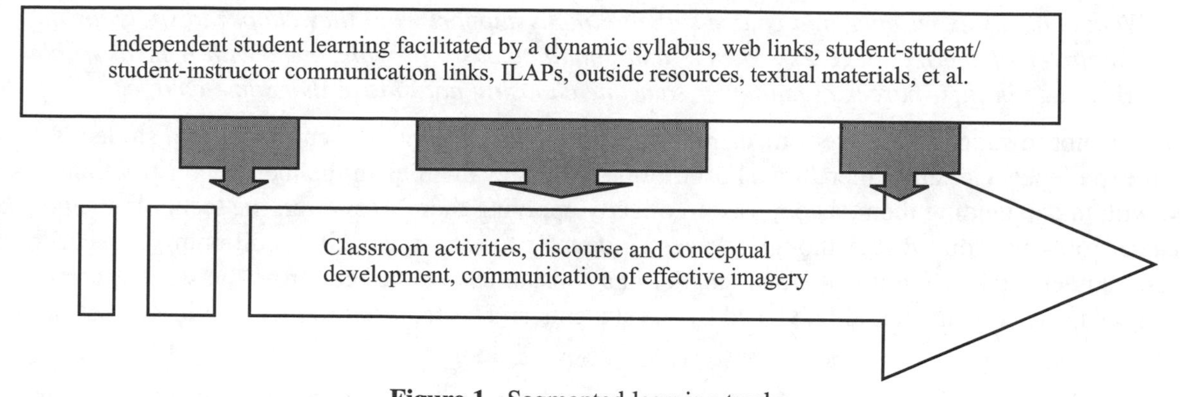

Figure 1. Segmented learning tracks

educators resort to creating templates and other "jump start" items to compensate for the learning curve in hopes that students will react by moving the task associated with it higher in priority. The ironic aspect of this observation is that the real learning experience is frequently found in the design, creation, and testing of these templates and not the simple execution of them.

In light of the previous observations, it seems that technology has the potential to free up class time by enabling material to be segmented into at least two tracks (see Figure 1). One track contains tasks and topical material that a student should be required to learn on their own. Conceptually, one might identify the topics in this track as those in which the student's own perspective characterizes the principal value-added dimension. In other words, their unique background of experience provides a depth and richness to the learning process that the homogenous environment of the classroom cannot achieve. Technology has a major role to play in this track because it affords educators with an opportunity to establish a "presence" in the learning environment largely crafted by the student, at their time and place of choice.

A second track then contains tasks and topical material that possesses subtleties best revealed through active discourse. In this track, it is the educator's perspective that, either through experience or education, defines the value-added dimension of learning. Topics that fall into this category are quite possibly those which are effectively learned only through the expenditure of time and effort and are, one might posit, the very reason that educators exist in the first place. Within the activities associated with this track, educators adopt the roles of mentor, learning guide, facilitator, and assessor, providing feedback and options to students that allow them to discover relationships and logical connections between conceptual abstractions that constitute true learning. Educators in this environment would find themselves frequently drawing pictures, talking with their hands, playing with technology, and performing various hands-on actions that attempt to craft consistent, mathematically sound imagery in the minds of students. The technology chosen to complement student learning in this track has the potential to be very different from that adopted to facilitate the independent student learning described earlier.

Technology's Affect on What We Teach

In an academic setting, there appears to be a curious tendency to not separate faculty and student perspectives when considering the question of integrating technology into a curriculum. By and large however, faculty perspectives on content and curriculum center on management issues: creating efficient methods for content delivery and course design, achieving the greatest "bang for the buck" in terms of effort expended for results achieved, developing accurate assessment methodologies, integrating and coordinating curriculum across departments, and so on.

On the other hand, the student perspective on what we teach that seems to have evolved is rather simply stated:

What students experience in a course should directly support what they choose to study for the remainder of their college experience, and simultaneously provide them with a measurable advantage in their career pursuit over someone choosing not to take that same course.

This is not to suggest that every mathematical topic be pragmatically measurable in a student's educational experience. Certainly there are a host of topics whose value lies in the mental constructions associated with understanding them, as opposed to directly applying them. However, the technology and mathematical tools introduced, the thought processes developed, and the tacit conditioning induced by the courses students take should matter, and matter in a substantial way as seen from the student perspective. Simply put, the content should be valued by the students and be required to meet a minimal set of objectives:

- It should be *relevant* to their worldview *now*, giving them a marked and measurable advantage in other courses they are taking, and set them up for success later.
- The focus should be the *problem*, not the mathematical tool. As such, it should make them better *problem solvers* in the most general definition of this term, not better tool manipulators. When the actual problem is the focus, the question of topical relevance is answered naturally.
- It should assist them in becoming discriminating users of information. They should develop a healthy skepticism when confronted with information and be able to recognize bad information when they see it. This is an especially desirable objective given the plethora of Internet resources and the lack of quality control structures imposed on them.
- It should equip them to be better able to organize their inquisition, extract key elements of information, and draw supportable and logical conclusions about increasingly complicated and sophisticated real-world problems. Hence, what we teach should condition them to ask the right type of questions to accomplish these tasks.
- Lastly, it should help students to recognize when they are truly looking at a hard problem, as opposed to what their limited experience tells them is hard because of their unfamiliarity with the course material or the task at hand. Some problems when confronted appear difficult but yield easily to certain mathematical methods (e.g., large-scale data analysis), whereas others appear deceptively simple but steadfastly resist any and all attempts to conquer them (e.g., traveling salesman problem).

The common ground between the two perspectives dictates that we should teach what is in the students' best interest, both long and short term. The courses and programs we design, the technology we adopt, and the pedagogical methods we use should focus on developing in the students the mental constructs and abilities they need to succeed in an increasingly technologically sophisticated world. The organization of these courses and programs should not be bound by the historically defined organization of mathematical topics as expounded in the current generation of textbooks unless such a textbook closely aligns with the desired objective. This implies that the baseline of what we teach should be constructed with a focus on mathematical ideas and abstraction that need development and not with a particular course currently in existence. Instead of examining the content of calculus under the assumption that calculus has to be the first course students should take in college, we begin such a design process by identifying and sequencing a progression of ideas and experiences that students must experience to achieve the goal stated earlier.

This enables, for example, elements of graph theory to be introduced at the onset because, in the progression of ideas that contribute to students understanding how mathematics applies to everyday life, graph theory contains objects that are in one-to-one correspondence with what students observe in life. They see a city, they draw a node; they see a major highway, they draw an edge; they know the distance between cities, they label the edge; how much material to ship from city i to city j, they label an x_{ij} on the edge, and so forth. This direct construction leads naturally to a mathematical representation to which solution algorithms can be applied, and the results of this model can be directly translated back to the real-world problem. At least one dimension of student growth can be directly assessed in such a design simply by introducing more complex problems for the students to consider over time. These complex problems would have the characteristic of requiring increased levels of abstraction to formulate the associated mathematical model. A similar growth component could be included for modeling situations dealing with data by increasing size and "dirtiness" of the data generated by the problem. The specific problems introduced would have an interdisciplinary focus that could also be tailored to the composition of academic majors in such a course. Thanks in a large part to technology, the mathematics curriculum of the future has the ability to be more adaptable than ever before, enabling creative educators to craft mathematics educational experiences custom tailored to meet the needs of the current student population.

A Unifying Framework

Given all of the above, we can suggest a unifying conceptual framework involving technology within which one might intelligently seek answers to the content question. Technology plays a role in such a structure as the principal workbench for faculty and students alike. Its main function is to be able to rapidly present various alternative perspectives on ideas and material with the hope that one of these perspectives will appeal to the student. A senior vice-president of a major technology company, who also happens to be a science fiction writer, once stated that "[he] didn't see anything extraordinary about mental telepathy." With each story he wrote his challenge was essentially the same: transmit the image of a particular scene he had in his mind into the mind of a person potentially thousands of miles away, who possesses a dramatically different life experience, and perhaps a very different world view. And, he had to accomplish this using only a judicious selection of words and phrases. This idea of communicating imagery is exactly what we employ in what follows.

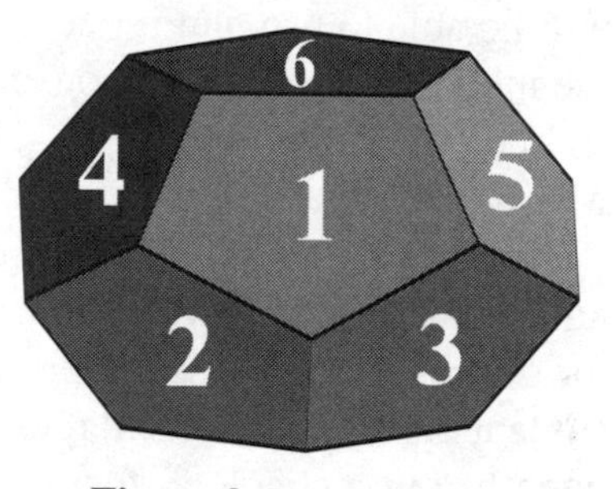

Figure 2. Imagery facets

Imagine a student sitting in the middle of a polyhedron such as the one displayed in Figure 2. Now imagine that we, as educators, exist outside the polyhedron, and each of the facets correspond to significantly different ways of representing a particular mathematical topic. Technology has increased both the number of facets at our disposal and the speed at which these facets can be created, presented, recalled, and manipulated. The educational challenge that remains despite technology's contribution is that the inside surface of each of these facets (the student's perspective) is almost assuredly not the same as the outside surface (that educator's perspective). Instead, there is a non-uniform probability of match associated with each facet whose distribution depends solely upon the student sitting inside the polyhedron. Technology should affect how we teach by better equipping us to cycle through the various facets in a decreasing order of match probability. Our motivation in doing so as a mentor, learning guide, facilitator, or assessor, is to assist a student in formulating a representation of the material that, if it does not match ours, at least is consistent with ours in the sense that it is a correct representation of the abstract structure associated with a particular mathematical topic. In mathematical terminology, it is isomorphic to the imagery we possess, that has proven over time to be valuable, relevant and mathematically correct in representing the real world.

Strengths and Weaknesses of Technology

At our institution, we have progressed through various experimental phases culminating in the situation we presently find ourselves in. Although each phase provided clues as to how best to adapt our technology strategy to achieve our academic goals, none have illuminated a globally optimal solution in light of the various dimensions successfully integrating technology into a curriculum involves. Most assuredly, not all choices have been met with unbridled enthusiasm by either the student body or the faculty, but even these failures have provided valuable lessons. For example, dictating that "every instructor will bring a computer to the classroom" in the late 1980s didn't work. The portability afforded by computers-on-a-cart was overshadowed by the logistics of maintaining such a fleet. Additionally, at that time only about 5–10% of our faculty were familiar enough with computing or possessed sufficient pedagogical insights to use this capability effectively. The remainder fell into the "pause and demonstrate mode" described earlier. In our current state, we have evolved to permanent installation of computers and projection devices in our classrooms that are all linked to a local area network. And, although the convenience afforded by such a setup has dramatically increased the number of faculty willing to experiment with using technology to augment learning, we have not yet solved the logistic challenge of maintaining up-to-date software and hardware without a significant financial commitment.

After a while, we began to recognize that our institution needed a unifying philosophy that enabled us to examine an emerging technology and be able to state: "It's a great technology that someone, somewhere will be able to use, but it doesn't fit within our learning environment framework." This has been perhaps the most important revelation emerging from our experimentation because it prevents us from investing in technology for technology's sake. This realization motivated the development of our learning model, which in turn gave us a clearer picture of the role technology could and should play within our curriculum.

Our current learning model recognizes that the greatest proportion of student learning occurs in a situation in which they have ample time for exploration, discovery, reflection and iteration. This is clearly not the 55 minutes of classroom time allotted for our mathematics classes. Instead, this time occurs when the students are back in their rooms since 100% of our student population is resident on-campus. Consequently, our technology focus has been to equip each student desktop with the most powerful computing platform available, connecting it to an unrestricted conduit to information on the Internet. Student portability of these resources, which points toward notebook computers and the like, has not emerged as a key characteristic of this design. Although this situation could change as time evolves and wireless computing becomes a reasonable alternative to hard wiring of networks, our current model implies that our technology investments not pursue portable technologies despite the wonderful capabilities they provide.

Two observations directly result from our technology-enhanced learning environment. The Internet affords a natural counter to student tendencies to perceive their USMA education as physically isolated. Foreign language classes can connect to distant newspapers to view the latest cultural events occurring in the Baltic region, France, Spain, and others. Political sciences and sociology classes link to many sites around the globe to provide supporting material for their studies. Outside research laboratories, government and civilian organizations, and professional societies all have a footprint in our mathematics courses. Students visiting these supporting sites preview potential career fields as well as obtain augmentation sources for current course work. As educators, we no longer are the sole source of information for our students, and neither is the course textbook.

Looking inward, technology holds great potential to counter student tendencies to view their academic courses as being modular, mutually exclusive experiences. Technology has grown to complement our interdisciplinary efforts over time, especially with regard to the creation and use of Integrated Lively Applications (ILAPS) and cross-departmental teaching. Moreover, technology allows us to actively link syllabi across the web, thus facilitating cross-departmental awareness of course content and teaching methods, enhancing research, and making real progress beyond rhetoric towards achieving a coordinated curriculum.

Our focus in the classroom is on the human dimension of learning: engaging in discourse, exploring conceptual ideas, conducting student presentations, and helping students to form efficient frameworks within which to view course material. The resulting increased interaction with students greatly complements our institutional objective to mentor the development of future leaders. Even within this human-centric framework, technology provides a practical workbench in the classroom to communicate key representations and be able to reach out from the classroom to obtain the right resource at the right time.

The drawbacks to our particular commitment to technology are few, but worth noting. We committed to a lifetime of maintenance activities involving both hardware and software upgrades on a periodic basis.

These are two sliding scales whose timing of events is neither synchronized nor convenient. We standardized the suite of software used by staff, faculty and students, subsequently finding our institution dependent upon single companies and their licensing nuances. We came face-to-face with reality: critical computing jobs are hard to fill, tough to keep filled once you fill them, and are a never-ending source of system weakness.

From a user's perspective, support for educational technologies is defined by needed response times, and these tend not to follow organizational charts. When response times exceed tolerance, departments create in-house expertise to facilitate effective support. Departmental expertise in computing technologies waxes and wanes over time. This imposes an additional administrative burden to maintain a continuous flow of talent. Faculty must also be cautious to not develop an over-dependency on this technology functioning correctly, as computer crashes, network switch failures and general software conflicts occur on a schedule apparently set by Murphy's Law. Our students have a diminished time window of tolerance for technology in that, if they do not get the results they expect (perhaps unrealistically) from a particular application, or if it has a relatively steep learning curve, they are quick to dismiss it. This impatience appears to be related to students' expectation of technological efficiency noted earlier.

Students and faculty that have invested the time and effort to experiment within this technology structure become enthusiastic advocates. The less adventurous manage to "sit and wait" on the sidelines, and the naysayers continue to attach their opinions to a small number of unfortunate data points, concluding that the marriage of education and technology will eventually end with divorce; it's just a matter of time.

Conclusion

A lesson that we are repeatedly taught through mathematics is that a solution is sometimes obtainable only when one artificially expands the obvious dimensions of the problem, thus providing a pressure-release of sorts on a solution methodology (e.g., Two phase method in Linear Programming, or homotopy methods in optimization). The eventual solution to effectively incorporating technology into education may be one that artificially (or not) reverses the fundamental roles of who is teaching and who is learning. Perhaps it does away with these roles entirely, making education a two-way cooperative venture in which educators have equivalent expectations of learning from the younger generation as the students have from the experienced generation of the resident faculty. Any curriculum adopted within such an environment must result from a dynamic, adaptable strategy that embraces change, not one that looks for stability in static components. If mathematics is the language of science and science changes with the times, then it is logical to expect that the structure of our language and the way we teach it must also evolve if it is not to suffer the same fate as Manx, the dead language of Turkey. Ultimately, change will come; if not brought about by this generation of mathematicians, then the next.

Towards a Framework for Interdisciplinarity

Lee L. Zia*
National Science Foundation

Abstract. We propose a yearlong integrated course of study emphasizing two themes: linear models in their own right, and linearity as a tool to help understand nonlinear phenomena (the process of linearization). By reorganizing mathematical content with an emphasis on contextual learning, the curriculum model we propose seeks to take advantage of the natural synergy of these three core mathematical subjects and their place within the broader context of second-year science, mathematics, engineering, and technology education.

Introduction

Linear algebra, ordinary differential equations, and multi-variable calculus hold an important position collectively within the standard second-year curriculum for many undergraduate science, technology, engineering, and mathematics (STEM) students, particularly those in engineering and the physical and applied mathematical sciences. Although many natural interconnections exist among these three subjects—with ideas and techniques from each in constant use in the others—their traditional organization and "delivery" as distinct courses can create artificial barriers among the subjects in the minds of the students. If we are to break down boundaries between disciplines for student learning, perhaps we should first break down subject boundaries within our own discipline. In this paper we explore an organizing framework consisting of two primary themes and several supporting conceptual axes, which we believe can promote an interdisciplinary approach to teaching and learning the central ideas from these three subjects.

Current Setting

Recognizing the centrality of the three subjects mentioned above, the importance of using motivating applications, and the opportunities for new pedagogical approaches afforded by advances in computational technology, the mathematical sciences community has been engaged over a number of years in various efforts to improve the teaching of these courses and student learning of this subject matter. For example, faculty enhancement workshops such as *Computer Aided Instruction in Linear Algebra and Ordinary Differential Equations* in 1990 [29] and the ATLAST [14] and CODEE consortia [6] both begun in 1992, have engaged large numbers of faculty across the country in rethinking their pedagogical approaches. More

* The views expressed in this paper are entirely those of the author, and do not reflect an official NSF position.

recently, a number of curriculum development projects have produced new teaching and learning materials for these three subjects. These include a MathCAD based laboratory approach to linear algebra by Porter and Hill [22], the *Internet Differential Equations Activities* project [15] and the text from the Boston University *Dynamical Systems* project [4] in differential equations, and the Calculus Consortium at Harvard's text *Multivariable Calculus* [16] and Cheung's work at Boston College with *Maple* and multivariable calculus [7]. Some of these efforts have led in turn to "second generation" faculty enhancement workshops and conferences as part of the dissemination work of these various projects. Additional projects and innovations are listed in the references.

While much progress has been made, several important challenges still exist, which argue for a conceptual and contextual unification of subject matter. Firstly, depending on a student's major requirements, the sequence in which students enroll in these courses can be quite varied; and institutions differ greatly in the frequency and the order in which these courses are offered. Other departments also teach versions of these courses for their own majors under slightly different names. For example, vector calculus courses for engineers are often taught under the name "Engineering Analysis". While this cafeteria plan of courses may maximize choice, it also encourages students' conceptual understanding to remain disconnected. Subject matter is "covered" repetitively, but not *revisited* in a coherent way. Opportunities to take advantage of and reinforce the synergistic interconnections among these three important mathematical subjects are lost.

Secondly, opportunities to make explicit linkages with application areas encountered by students in their major-specific courses are not generally exploited. Often these are courses in which they are either concurrently enrolled as second-year students or will be enrolled as third-year students, e.g. electromagnetism, statics, circuit theory, or systems and control. This practice of fragmentation ignores the larger curricular context in which the courses are being offered.

Finally, while not a universal situation, the fact remains that many students take these courses in large enrollment settings that have generally not proven to be conducive to optimizing student learning.[1] In this paper we will not address explicitly the issues involved in restructuring student learning environments (see for example [5,8,9,21]). However, the subject is a rich and important one on which many continue to work.[2]

Conceptual Framework

> "*All exact science is dominated by the idea of approximation.*" — Bertrand Russell[3]

We propose a yearlong integrated course of study emphasizing two themes: linear models in their own right, and linearity as a tool to help understand non-linear phenomena (the process of linearization). These dual themes would offer a way to unify the "standard topics" contained in the typical three-course collection of linear algebra, differential equations, and multivariable calculus, and to promote the connected learning of their core ideas. While the total number of "contact hours" for a course of study of this type would be the same as the three separate courses combined, we believe that the *whole can be greater than the sum of its parts*. Typically, for any given unit of time within a term, students are constrained to an equal

[1] Characteristic of these settings is the concatenation of large-lecture sections taught by faculty with smaller recitation sections taught by graduate students or adjuncts. Students face different instructional approaches and styles in this practice, and are also often with different classmates depending on the learning venue, further fragmenting their learning experience.

[2] The integration of formal lecture sections, recitation and/or discussion sections into a "studio" or laboratory setting coordinates and reinforces opportunities for student learning and understanding. In recent years, RPI has developed successful studio approaches to teaching and learning, e.g., Studio Physics [28] and Studio Calculus [10,11].

[3] B. Russell, "The Scientific Outlook", p. 63, *The Free Press*, Glencoe, IL, 1931.

distribution of formal classroom time in each of the three courses (or worse, only one or two of the courses with the others taken in another term!). Variable sequencing of the courses further contributes to creating a set of disconnected learning experiences.

The model we propose seeks to break out of this administrative confine, by creating a learning structure that embraces the rich interconnectedness of these three subject areas and enables students to learn the core ideas in a *concurrent* manner. Each of the three single-term courses typically carries three or four credits, with four to five contact hours each week. In place of this structure consider one in which there are six contact hours weekly throughout a yearlong (two-semester) sequence. To maximize active student engagement with concepts through a laboratory learning environment, a combination of one- and two-hour class blocks totaling six hours could be scheduled weekly. In any given week of a term, the relative amount of "coverage" of material usually identified with linear algebra, ordinary differential equations, or multivariable calculus can vary, and this distribution can change from week to week. At the end of such a course of study, we believe the cumulative effect can be at least equivalent to that of an existing system in terms of mere "coverage" of standard topics, but *much greater* in terms of student learning, as a result of the coordinated context in which the learning takes place.

Several projects are in fact working towards this vision. For example, the Linearity I and II project (Black et al, see [3]) is experimenting with the structural changes and the use of "mini-projects" from science and engineering, while the Coordinated Curriculum Library project (Moore et al, see [20]) and the Connected Curriculum project (Wattenberg et al, see [27]) are developing materials and laboratory modules.

Supporting the dual themes of linear models and the linearization process, we envision several axes representing different modeling perspectives that collectively define a conceptual space into which projects, experiments, and other learning materials and tools can be placed. One axis consists of a one-dimensional to multi-dimensional perspective, a second axis represents the interplay between discrete and continuous models, and the third axis captures the contrast between deterministic and stochastic phenomena and/or assumptions.

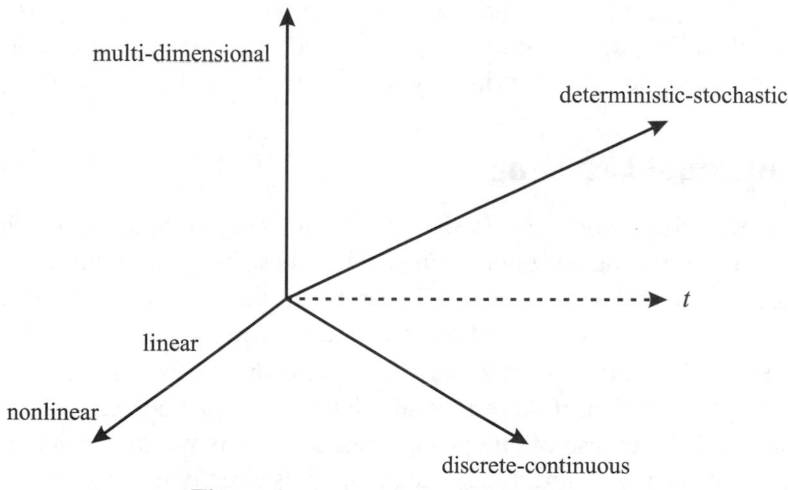

Figure 1. A conceptual framework

A Prototype Scope and Sequence

If the first-year curriculum in mathematics aims to develop a rich collection of *ways of knowing* a single-valued function of one variable, then the second-year should develop a similar understanding for single- and vector-valued functions of several variables. The most fundamental of these functions are the linear ones, and the presence of multiple variables (both input and output) argues naturally for taking up linear systems from the outset. Thus, we envision the year sequence beginning with linear systems of equations and developing the basic language of linear algebra: vectors, vector operations, superposition, etc.

While in the first two or three weeks the distribution of coverage of standard topics would be weighted significantly towards linear algebra, this emphasis could be balanced by an early introduction to the second major theme of the course: the use of linear approximations. In both cases these introductions can be naturally connected to and complemented by realistic laboratory experiments, either brought directly into the mathematics classroom or investigated concurrently in allied departments. For example, truss problems such as those described below in greater detail provide a natural way to discuss linear systems of equations. Likewise, initial investigations of surfaces and level curves, partial derivatives and gradients, and tangent plane approximations could be linked to electrostatics, where students could observe and construct equipotential lines on a charged plate. Elementary fluid flow examples could also provide useful context.

In our hypothetical model, the first term of the sequence would deal primarily with static models of science and engineering phenomena that find expression as purely algebraic equations. A particularly interesting and value-added feature afforded by this approach is that the data fitting issues that arise naturally in analyzing experimental observations lead directly to least squares problems. This of course leads to orthogonality, Gram-Schmidt orthogonalization, and the QR algorithm. Moreover, the groundwork is laid for more advanced model validation issues encountered in upper-division courses. During the second term of the sequence, we see the program of study changing to one largely dominated by consideration of dynamic models and science and engineering phenomena which express themselves through dynamical systems of equations. It would be especially valuable to consider examples that begin as a static model, and can later be revisited as a dynamic one.

Within the framework depicted in Figure 1, it is possible to bring out the important interplay between the arithmetic and algebra of discrete observations and the calculus of the infinite. For example, a natural introduction to dynamical phenomena and models is provided by examination of difference equations $x_{n+1} = Ax_n$. This approach capitalizes on the early grounding in the language of linear systems we propose. Eigenvalue-eigenvector analysis can be motivated by numerical computational experiments and visual representation of the iterations, which reveal the dominant eigenvalue and associated eigendirection.[4] The concepts of equilibrium and stability arise naturally along with the use of linear approximations of the multivariable functions that define the vector fields. Subsequently these ideas are revisited in the form of differential equations and dynamic (time-varying) models of science and engineering applications.

Promoting Contextual Learning

We believe the conceptual framework described above can provide multiple entry points from which interaction with science and engineering colleagues can proceed to establish and reinforce connections with concurrent, core disciplinary courses such as statics, linear circuit theory, dynamics, structures, electricity and magnetism, fluid mechanics, and systems and control. Indeed, there is emerging evidence (see for example [17–19,23]) that conceptual learning can be positively impacted by the use of physical models and hands-on experiments. While implementation of these cross-disciplinary interactions can take the form of guest lectures, demonstrations, and shared use of equipment, we believe it is worth considering the systematic and systemic incorporation of such experiment-based learning tools directly into the mathematics classroom.

For the past several years innovative faculty have begun to take advantage of the ready availability of simple hand-held Calculator Based Laboratory (CBL) devices from Texas Instruments, and more recently Palm Pilot platforms [26] have even begun to make their way into middle school classrooms. We describe four examples that are chosen to help illustrate core mathematical ideas and provide opportunities to make explicit connections to concepts that students encounter in their courses used to satisfy major requirements.

[4] This is of course the basic idea behind power methods for numerical eigenvalue calculations.

Example 1: Structures in equilibrium

Planar and three-dimensional trusses are considered, both constructed for illustrative purposes and taken from the real world.[5] The related subject matter in mechanical engineering and civil engineering as well is the topic of *statics*. Strain gauges record the tension in the frame members as a result of loads applied at different nodes of the structure. A linear system of equations $\boldsymbol{Ax} = \boldsymbol{b}$ relates the vector of displacements in the frame members to the vector of applied forces at the nodes. The matrix $\boldsymbol{A}$ is known as the stiffness matrix.

This case provides an example of a linear system of equations and an introduction to Gaussian elimination as the method of solution. Linearity or the principle of superposition can also be experienced first-hand by students as they observe the effect of applying different vectors of loads and measuring that the output of the sum (of inputs) is the sum of the outputs (from each input). The $\boldsymbol{LU}$ decomposition[6] that records the steps of Gaussian elimination is motivated by the problem of needing to determine the responses of the truss due to multiple loading vectors (different right-hand side vectors $\boldsymbol{b}$). A particularly attractive feature of this example is that the two methods of analysis used by engineers, the so-called method of joints and the method of sections are in fact mathematically dual formulations of the problem.

Many points of departure are possible from this example including the investigation of nonlinear stress-strain relationships and a subsequent linearization to understand local behavior, and the investigation of ill-conditioning via consideration of a nearly statically indeterminate system. The latter investigation has particular importance, since the increased use of powerful software packages for numerical computation argues for increased attention to understanding issues of numerical stability.

Example 2: Analysis of electric circuits

Basic examples of linear electric circuits are certainly an important component of the electrical engineering and physics curriculum. The connectivity matrix (or edge-node incidence matrix) describes the topology of the electrical network. Kirchhoff's Laws and Ohm's Law combine to yield linear systems of equations relating potentials and currents to applied voltages and current sources. Measurement devices can allow students to observe first-hand the linear input/output relations embodied in these circuits.

As in the preceding example, this case presents another source of real-world matrices and linear systems of equations.[7] In fact, Kirchhoff's Voltage and Current Laws find equivalent mathematical expressions in terms of descriptions of the column space of the connectivity matrix and its left nullspace, respectively. Key mathematical concepts such as orthogonality and the interrelationship of the four fundamental subspaces associated with a matrix arise concretely in terms of the physical context (see [25] for example). Furthermore, a discussion of linear independence and basis can be made explicit through reference to loops in the graph and to determining the number of these loops that are independent. In terms of the discrete-continuous conceptual axis alluded to above, the edge-node incidence matrix can be viewed as a discrete approximation to the derivative operator. Then the fact that its nullspace consists of the one-dimensional subspace of constant vectors coincides with anti-differentiation only being determined up to the set of constant functions.

Again, there are many points of departure for subsequent re-visitation afforded by use of this case. Introduction of capacitors and inductors to the electric circuit moves the problem into the realm of dynamic or time-varying models, i.e., differential equations. But preserving the fundamental linear nature of the prob-

---

[5] For example, members of the Department of Mechanical Engineering at Rensselaer Polytechnic Institute have used bicycle frames for similar purposes.

[6] The more general matrix factorization is $\boldsymbol{PA} = \boldsymbol{LU}$ to accommodate row interchanges.

[7] Changing applied voltages in these circuits is analogous to changing applied forces in the truss problems, with both problems providing additional entry points to a discussion of stability of matrix calculations and conditioning.

lem yields another opportunity to encounter the study of eigenvalues and eigenvectors. Further on, nonlinear circuits could be considered, for example students could experiment with a van der Pol oscillator, allowing motivation for the qualitative analysis of the phase plane. The particular consideration of a van der Pol oscillator would also allow investigation of bifurcation phenomenon and an opportunity to motivate the use of linearization to gain insight into local system dynamics.

Example 3: Oscillations and periodic phenomena

An air track with a collection of masses and spring is the physical system considered. This application is representative of subject matter in mechanical engineering, electrical engineering, and physics relating to the broad area of *dynamics*. Newton's Second Law and Hooke's Law combine to yield a second-order linear system of differential equations

$$\frac{d^2x}{dt^2} = Ax \; .$$

This example again provides a concrete example of a linear system of equations. For a two-mass, three-spring system (with equal masses and equal spring constants), the corresponding matrix

$$\boldsymbol{A} = \begin{bmatrix} -2 & 1 \\ 1 & -2 \end{bmatrix}$$

has eigenvalues which yield the frequencies of oscillations of the so-called "fast" and "slow" modes of oscillation.[8] Furthermore the eigenvectors of $\boldsymbol{A}$ correspond to the two initial conditions (the vector of initial displacements, with zero velocity) that produce exactly these fundamental modes of oscillation. One mode begins with the two masses displaced an equal amount in the same direction, $\boldsymbol{x}(\boldsymbol{0}) = [\boldsymbol{1},\boldsymbol{1}]$ and the second mode begins with the two masses displaced equal amounts in the opposite direction, $\boldsymbol{x}(\boldsymbol{0}) = [\boldsymbol{1},\boldsymbol{-1}]$.[9] For further investigation, students could increase the number of masses and springs, consider the effect of different masses and different spring constants (qualitative dependence on parameters), include frictional effects in their model, and consider nonlinear effects as well. A particularly interesting avenue to follow is to consider the limiting case of an infinite number of masses and springs (each of smaller length), a model that could lead to a discussion of the wave equation and conservation principles.

Example 4: ExFluid behavior

Experimental apparatus can also be constructed to illustrate fluid mechanics concepts that are particularly germane to students in mechanical engineering and physics. The Hele-Shaw cell consists of two clear plastic plates separated by a thin space into which a fluid may be injected through various input holes. The narrow spacing confines the fluid to be essentially two-dimensional. Here we envision that the use of a real physical fluid can help illustrate a vector field and motivate concepts such as its divergence and curl. To that end, students could inject colored dye into the fluid to help mark the flow field, or alternatively, small particles such as aluminum chips can be used. If the fluid velocity is slow enough, actual measurements can be taken. Incompressibility can be felt and mathematically verified $\mathbf{div}\ \mathbf{v} = 0$.

[8] In fact the frequencies are the square roots of $-\lambda_j$, the negative of the eigenvalues.

[9] The author has used this experimental device numerous times in both linear algebra and ordinary differential equations courses. Students count the number of oscillations for the different modes in a ten-second interval and then compare their ratio. This procedure usually yields one decimal place accuracy to the theoretical value, much to the surprise and pleasure of the students. Students also report overwhelmingly that such an experiment contributes to their understanding and helps connect their mathematics courses to their other science and engineering courses.

Conclusion

By reorganizing mathematical content with an emphasis on contextual learning, the model we propose seeks to take advantage of the natural synergy of these three core mathematical subjects and their place within the broader context of second-year STEM education. In combination with a restructured learning environment, this approach can offer a vehicle for bringing thematic and pedagogical coherence to much of the second-year STEM curriculum. This in turn provides a unified framework for learning that is applicable to a significant number of STEM students. While we believe that many institutions will find value in this approach, it may hold special appeal to institutions that enroll large numbers of engineering and/or applied science students. In particular, the "coherency and efficiency of coverage" afforded by the framework may help address an overcrowded engineering curriculum. For engineering schools in particular this approach is consistent with new ABET certification requirements. It is also interesting to speculate about the application of this framework to linear algebra, multi-variable calculus and probability and statistics in support of a curriculum that reflects a focus on environmental/civil engineering, chemical engineering, and earth science.

References

1. Banchoff, T., *Interactive Electronic Third Semester Calculus Laboratory Materials for Personal Computers*, Brown University, NSF/DUE-9450721.
2. Baxter-Hastings, N., *Using a Workshop Approach and Real Applications to Teach Calculus with Review Courses*, Dickinson College, NSF/DUE-9450746.
3. Black, K., et al, *Linearity I and II: Connected Learning of Ordinary Differential Equations, Linear Algebra, and Multi-variable Calculus*, NSF/DUE-9752650.
4. Blanchard, P., Devaney R., et al., 1998. *Differential Equations*, Brooks/Cole.
5. Bookman, J. and Friedman, C.P., 1994. A Comparison of the Problem Solving Performance of Students in Lab Based and Traditional Calculus, in *Research in Collegiate Mathematics Education I*, CBMS: Issues in Mathematics Education, vol. 4, Providence: American Mathematical Society.
6. Borrelli, R. and Coleman, C., *Computer Experiments in Differential Equations*, Harvey Mudd College, NSF/DUE-9154300.
7. Cheung, C.K., 1998. *Multivariable Calculus with Maple*, New Tork: John Wiley and Sons.
8. Davidson, N., ed., 1990. *Cooperative Learning in Mathematics: A Handbook for Teachers*, Reading, MA: Addison-Wesley.
9. Dubinsky, Ed., *Cooperative Learning in Undergraduate Mathematics*, Georgia State University, NSF/DUE-9653383.
10. Ecker, J., 1996. *Studio Calculus*, New York: Harper Collins.
11. ———, 1996. *Instructors Resource Manual for Studio Calculus*, Reading, MA: Addison-Wesley.
12. Friedman, A. and Littman, W., 1994. *Industrial Mathematics: A Course in Solving Real-World Problems*, Philadelphia: SIAM Press.
13. Laws, P., Calculus-based Physics Without Lectures, *Physics Today*, December 1991, 24.
14. Leon, S.J., *Project ATLAST*, Univ. of Massachusetts–Dartmouth, NSF/DUE-9154149.
15. Lofaro, T., *Internet Differential Equations Activities*, Washington State University, NSF/DUE-9555228.
16. McCallum, W., Hughes-Hallett, D., Gleason, A.M., et al., 1997. *Multivariable Calculus*, New York: John Wiley and Sons.
17. Monk, G.S., Students' Understanding of a Function Given by a Physical Model. In Dubinsky, E. and Harel, G. (eds.), *The Concept of Function: Aspects of Epistemology and Pedagogy*, MAA Notes 25, Mathematical Association of America.

18. Monk, G.S. and Nemirovsky, R., The Case of Dan: Student Construction of a Functional Situation through Visual Attributes, in *Research in Collegiate Mathematics Education I*, CBMS: Issues in Mathematics Education, vol. 4, American Mathematical Society, Providence, RI, 1994.

19. Monk, G.S. and Nemirovsky, R., *CalcLab: A Hands-on Learning Environment for Enriching Students' Understanding of Calculus*, TERC and the University of Washington, NSF/DUE-9653068.

20. Moore, L., et al, *The Coordinated Curriculum Library*, NSF/DUE- 9752421.

21. Park, K. and Travers, K., A Comparative Study of a Computer-Based and a Standard College First-Year Calculus Course, in "Research in Collegiate Mathematics Education II.", *CBMS: Issues in Mathematics Education*, vol. 6, American Mathematical Society, Providence, RI, 1996.

22. Porter, G.J. and Hill, D.R., 1996. *Linear Algebra: A Laboratory Approach with MathCad*, New York: Springer-Verlag.

23. Russell, D., *Work Station For Undergraduate Laboratory In Applied Mathematics*, Virginia Polytechnic Institute and State University, NSF/DUE-9350736.

24. Smith, D.A., et al, (eds), 1988. *Computers and Mathematics: The Use of Computers in Undergraduate Instruction,* MAA Notes 9, The Mathematical Association of America.

25. Strang, G., 1993. *Introduction to Linear Algebra*, Wellesley-Cambridge Press.

26. Staudt, C., Probing Untested Ground: Young Students Learn to Use Handheld Computers, *Concord Consortium Newsletter*, The Concord Consortium, Concord, MA, Fall, 1999.

27. Wattenberg, F., "The Connected Curriculum Project", `www.math.montana.edu/~frankw`

28. Wilson, J.M., The CUPLE Physics Studio, *The Physics Teacher*, vol. 32, p. 518 (1994).

29. Zia, L.L. and Bechtell, H., *Computer Aided Instruction in Linear Algebra and Ordinary Differential Equations*, University of New Hampshire, NSF/USE-9054195.

Pedagogically Effective Use of Technology

Joseph D. Myers
United States Military Academy

Abstract. Student-accessible technology is now almost universally available and its use in our classrooms is increasing rapidly. In some quarters it has become almost the badge of reform to be known as a technology user. Our students are inundated with graphing calculators, CAS systems, and specialized software packages, each weighing in with performance claims in their respective niches. Is our plan pedagogically sound, and are we heading in a direction that best benefits our students? In this paper, we investigate issues surrounding technology tradeoffs, the uses of technology, and the potential future technology holds for mathematics education.

Introduction

Technology should change the set of skills and knowledge that we have traditionally professed as fundamental. Some skills are no longer important to most potential users of mathematics (e.g., most of our students) because technology is almost universally available that can execute them sufficiently well both within and outside of academia. Yet we are mysteriously drawn to dedicating valuable student contact time to teaching topics of this nature. Techniques of integration, drawing and graphing, root finding, and solutions to systems of equations, both linear and nonlinear, all fall into this category. Moreover, despite the fact that technology's ability to make nearly effortless numerical evaluations has greatly reduced the importance of trigonometric identities, we still employ them just because they are available and we (as teachers) know them. We often bemoan the fact that our students don't know the trigonometric functions at special angles, but why do we value that so much? 30° and 60° angles are largely artificial; they show up so rarely in real applications that maybe we should satisfy ourselves with numerical evaluation as needed, just as we do for 31° and 59° angles. Cross products and curls are easily evaluated with technology and there is no mathematical insight gained through calculating them by hand, yet we persist in teaching and testing these topics.

Accepting the fact that some skills such as those discussed above have diminished in importance, we must recognize that there is a corresponding set of technology skills that gain critical importance in response. These include skills in numerical evaluation, computer algebra, linear algebra, numerical computation (of roots, eigenvalues and eigenvectors, combinations and permutations, and statistical measures), and graphing and visualization. We have taken a very selective approach here; most of us teach a few random and personally convenient skills, but few of us have a comprehensive inventory of what students need or a plan for covering them all systematically. Reference [1] contains a sample plan of such skills, with representative realizations.

Some skills and knowledge remain important, and deserve added attention because of technology-induced atrophy. These include basic geometry, the algebra of polynomials, exponentials, and logarithms, the derivatives, integrals, graphs, and behaviors of the elementary functions, and the domains and ranges of the vector differential operators. Reference [2] contains a sample list of such skills.

Technology should allow students to realize that, at least on one facet, mathematics is an experimental science. Technology allows us to disprove conjectures in a straightforward manner, but it can also be used to illustrate that *proof by example* is support for beginning a proof even though it is not a proof. It means to do enough (varied) positive examples to convince oneself that a conjecture is very feasible, and that (coupled with our assurance) he or she really should believe it. What this step also does is motivate the need for proof; with so many examples under the student's belt and the willingness to believe it's true, they are now motivated to pursue the proof (if we have an audience that has either the need or an inclination in that direction).

In this vein, reform efforts should target places where deductive and analytic approaches have historically failed students to examine whether technology offers an effective surrogate. For example, deductive approaches to convergence of series are traditionally disasters for students, both in understanding and in execution. Replacing these with a series of experiments designed to lead students toward a visceral understanding of convergence tests might go a long way towards improving this situation. Consider the following example of how one might construct such an experiment.

Example 1. Guess at a test for convergence ($a_n > a_{n+1}$?). Given an intelligently selected collection of instructor-provided series, graph "enough" partial sums to tell if each is converging or diverging; can you use these to give evidence for or against your conjecture? Plot the ratio a_{k+1}/a_k; is there any relationship between convergence and the graph of this ratio over k? Does the harmonic series look convergent or divergent? Group its terms into packets such that the nth packet contains $2^n - 2^{n-1}$ terms and plot the sum of each packet; can you draw any conclusions? This approach is applicable to other traditional student problem areas, such as limits and continuity of multivariable functions.

We should move beyond using technology to demonstrate its applicability to "toy problems" and really start addressing problems more appropriate to their abilities. As technology proponents we often tout the ability to do more realistic problems and yet persist in teaching traditional skills and techniques on canonically easy problems, repeating those problems using technology to show how it is done and how technology makes it even easier. This is not to advocate the introduction of nonlinear equations, uglier integrands and more complicated functions just to demonstrate technology's worth. Rather, we should actually take advantage of the power available by refining our models to more closely reflect reality and show how solving refined models leads to refined solutions.

Which Technological Tool(s) to Use?

In general, it seems that the universality and general public familiarity of a given technology choice is inversely proportional to its power. One way to group various technology choices is by function. Most of our technology choices fall into the area of technology that helps us *do mathematics*, a few choices help us *demonstrate mathematics*, and a few help us to *communicate* as part of the learning process.

Doing Mathematics

Graphing calculators have proven to be economical and popular with the rising "Game Boy" generation. Their reasonable costs have made them nearly ubiquitous in high schools and as graduation presents. This situation can potentially provide a free and instant jump-start into college technology use. Their portability coupled with the availability of optional sensor-peripherals make these calculators ideal for turning a classroom into an instant laboratory.

One major disadvantage is the rapid pace of development; the market is so rich with quality choices in this area that compromises must be made in what will be used or required or in who will have to switch and learn a new calculator. Another disadvantage is in the power available and the ease of use to access it.

Visualization graphics are often mediocre and navigating the layers of menus required to execute some operations is arcane and tedious.

Besides being universally available and familiar to many, spreadsheets are ideally suited to many applications involving sequences, series, difference equations, discrete approximations to continuous operations, iteration, or visualization of the same. It is both an advantage and a disadvantage that they require the student to understand the mathematics well enough to be able to mentally flowchart what they are trying to accomplish and to construct and implement an algorithm that matches this logic. Being constrained to numerical, discrete applications, spreadsheets are often bypassed by mathematics faculty in favor of packages that are somewhat more general, yet are often favored by faculty in our partner departments.

Computer algebra systems (CASs) have proven to be a more popular choice than others because of their ability to handle a wide range of discrete, continuous, and visualization tasks. Inexpensive CASs are accessible to students, but make tradeoffs in power and generality. More powerful CASs are understandably more expensive, which often limits their use to laboratory settings and server installations rather than individual distribution to student populations. Non-intuitive syntax is a problem with almost all CASs, carrying with their use a relatively steep learning curve.

A fourth choice is one of the increasing numbers of specialized mathematics/engineering packages. Packages for linear algebra, ODEs, statistics, complex analysis, and other specialized areas are available for use in the appropriate courses. Advantages include a programming structure and command set that are tailored for the field of interest. This is also their chief disadvantage: the packages are seldom used or useful outside the one or two courses in which they appear.

Several schools have experimented with issuing notebook computers to students, either as a required purchase for all incoming students or as one-semester or one-year loans in order to conduct classroom studies of effectiveness [3]. The ability to turn any given lesson of any given class into a laboratory session holds great promise, as does the ability to have students actively involved with examples in class. Their inherent ability to transport work and programs between class, dorms, and trips weighs in their favor, while weight, security, durability, backup power in the classroom, and boot time in the classroom list among their disadvantages.

Demonstrating Mathematics

Web applets are an increasingly popular way to demonstrate to students such mathematical concepts as Riemann sums, secant and tangent lines at a point, the area interpretation and direction of a cross product, etc. These can be made available to large audiences at the individual's convenience via web access and offer an active and visual way for students to play with a concept. However, they require some amount of faculty creativity and programming expertise to fully exploit their potential.

Crafted into another mode, CASs can also be used to directly demonstrate mathematical concepts. By first creating customized application files and then posting or electronic mailing these to students, these CAS files require the student to simply execute them, thus allowing the student to change no more than a few parameters or functions to solve very complicated problems. This approach is popular with students because they require only a basic competence in the CAS to experience the mathematical point being made. One disadvantage of this technique is the dependence upon instructor time and ingenuity. Moreover, faculty members who have implemented this approach frequently worry that making too many of these CAS files publicly available to students might cause the students to overlook the underlying mathematical concepts the files were created to demonstrate in the first place.

Communicating

Email over campus networks is an increasingly common way to enhance learning for students. We increasingly send out to students administrative notices and requirements (which saves more classroom time for teaching and learning), night-before tips for what to key on and what to de-emphasize in a given night's reading assignment, immediate corrections or clarifications to something that happened in the class that just adjourned, and so on. As this medium matures, it seems that we move from the initial stage where we contact students, to an intermediate stage where we augment what we are saying to students, and finally to an advanced stage where most of a student's professors are inundating him or her with daily odds and ends.

A web presence, such as a personal or course webpage with pertinent course materials and important notices and updates posted, makes materials available to students without being overbearing to the point of student rejection. The effectiveness of this option depends on individual student web use habits and compulsiveness.

Matching Technology to Audience

This is one area that requires constant attention with an eye toward improvement. Faculty decisions in this vein tend toward sub-optimization, both as individual instructors and as one department within a larger academic institution. Technology to be used (if any) in a particular course is principally based upon what the faculty is comfortable with. We may secondarily consider what a course has historically used and if our predecessors have left any worksheets or materials that we can reuse. The result is a technological experience for students that has them learning (or attempting to learn) several new technology choices in several different courses in an often disjoint way, and then forgetting them as they are no longer required or used.

Adopting a more global perspective than one principally focused on mathematics courses might illuminate a more pedagogically sound technology choice for use in our programs, especially in our courses designed to support other departments and academic majors programs. Committing to one principal choice or to a few common choices and consistently using and building on these so that students can develop long term familiarity, faith in their own competency, and willingness to use technology on problems that arise in other courses or in open-ended contexts seems more logically sound than attempting to expose all students to all technologies.

Implications

Some believe that we are seeing the emergence of a new learning style among our students. Growing up in a technologically rich society may be making our current and future students more comfortable learning from electronic sources than from printed or hard-copy sources. When syllabus and lesson assignments are available both in hard copy and on the web, some students ignore the hard copy in the notebook on the side of their desk and look up the web version. Many students even ignore the texts on the shelf in front of them and instead surf to research and find information. The traditional feeling that "electronic is nice, but I need hard copy to study from" may soon no longer be the case for our students. A lifetime of exposure to technology could be changing attitudes about which medium is preferable.

Conclusion

There are many individual success stories concerning technology in the classroom. Graphing calculators and CAS use have exploded in the last ten years. There are many factors indigenous in these technologies that appear to work in our favor. Increasingly powerful technology choices continue to rapidly evolve to reasonable cost levels. Software interfaces continue to slowly emerge that employ syntax which may actually be intuitive to the new user. Our departments continue to attract new faculty who grew up with technology, and an increasing number of our faculty hold the sincere belief that technology can be a useful and illuminating vehicle for learning. An increasing number of faculty talk, write, and give and attend workshops about how to use technology in the classroom. Institutional support for classroom technology initiatives is almost surprisingly strong. In all, it appears that technology in the classroom is past the fad stage, and is a recent part of our pedagogy that is here to stay.

References

1. Department of Mathematical Sciences, 2002.*Core Mathematics at USMA*, West Point, NY, 29–32.

2. Department of Mathematical Sciences, 2002. *Core Mathematics at USMA*, West Point, NY, 25–28.

3. J. Loy, J. Myers, and C. Tappert, "Notebook Versus Desktop Computers for Cadets at West Point", *IEEE Transactions on Education*, Vol 39, No 4, Nov 1996, pp. 497–504.

Technology in the First Two Years of Collegiate Mathematics

Wade Ellis, Jr.
West Valley College

Abstract. We present several roles of technology and suggest various ways that technology could have a lasting and significant impact upon the quality of mathematics courses being taught in the first two years of collegiate mathematics. Overcoming some mathematicians' anxiety and reluctance to address applied problem solving so as to take full advantage of the opportunities remains a challenge for the future. However, we suggest that the inertia of change in the educational landscape makes applied problem solving a necessity, not simply for the creation of more interesting mathematics courses, but to better serve the needs of students and our partner disciplines.

Introduction

Technology affects the teaching and learning of mathematics in the first two years of collegiate mathematics in many ways. It is a platform for presenting ideas, an engine for restructuring mathematical course content, a tool for tutorial review, an Internet communication and data collection device, and a way of performing mathematical computations with numbers, symbols, and graphs.

The most obvious of these uses, as a platform for presenting ideas, merits passing comment that professors have historically used technology tools such as transparencies and overhead projectors poorly. Upon occasion, they have unwittingly caused temporary blindness in themselves while simultaneously obstructing the view of students directly in front of the overhead projector. With the introduction of new, optically sharp, lightweight computer projectors that can be mounted in the ceiling of a classroom and connected to a variety of devices controlled from a computer, the situation has improved substantially. An instructor can display marvelous class notes, simulations, animations, and computer-driven computations quickly and easily. Preparing presentations that take full advantage of this technology is time consuming but can be very effective.

An Engine for Restructuring Mathematical Course Content

Computer software that performs mathematical computations can be a driving force for reexamining the content and emphasis of our existing mathematics courses. Looking across disciplines we've seen how even low level technology has in the past changed physics courses. Students no longer spend valuable hours computing answers to a few challenging problems using paper and pencil or slide rules. Rather, they engage a greater number of problems having greater diversity because they perform the requisite computations with a scientific calculator or computer. The content of the courses may not have changed much, but the time students spend on critical thinking about physical principles has increased.

In mathematics, we have not yet fully addressed the difference between mathematics and mathematical computations. This can be seen most vividly in the importance that has historically been given to the computational gymnastics associated with the differential equations course that is the capstone course in mathematics for engineers. The wonderful applications that become accessible through a broader understanding of differential equations were frequently overlooked in the quest for increasingly elaborate computations to solve more and more specialized differential equations. That most differential equations could best be approached from a numerical, qualitative, and graphical standpoint is often lost in the effort to make sure students can compute quickly and accurately, something a computer is better suited to do.

Some mathematicians do not value the applications of mathematics and as a consequence do not give emphasis to them. Others like even less the inroads that computer-driven computations and graphics have made in the use of mathematics. Theory is an essential (if not *the* essential) ingredient of our discipline. However, most students do not major in mathematics and need to know the limitations that theoretical mathematics place upon the investigations of applied situations in the real world. If we only seldom address applied problems, then our students will not develop an ability to understand and use the lively interaction between theory and the real world. For example, students unfamiliar with the need for existence theorems in differential equations may begin to use numerical techniques to solve equations that have no solution or no unique solution.

That differential equations should come earlier and be treated in more depth has been foreshadowed by some of the curriculum reform efforts, but has not been fully implemented because computer symbolic manipulations have not yet been fully embraced by educators. There is perhaps a need to introduce difference equations early in the first course in calculus and to use them as a springboard to differential equations as has been done at the U.S. Military Academy. Perhaps the interaction between differential equations and difference equations in investigating real world situations could be exploited in such courses. Also, the requirement that power series be treated in a first course rather than multivariate calculus concepts should perhaps be revisited.

A Tool for Tutorial Review

There are now available many assessment and tutorial programs for various mathematics courses given at the college level. Infrequently, these programs are combined into a package that both assesses the student in some content area and then provides the appropriate tutorial topics to master the content area. ALEKS (Assessment and LEarning in Knowledge Spaces, `www.aleks.uci.edu`) is a National Science Foundation (NSF) funded cognitive sciences project that is now reaching the commercial stage that does just this. At the moment, it is only available on the Internet for beginning algebra, intermediate algebra, and geometry. Nevertheless, such a package of services for the student could improve the retention rate in calculus courses by providing students with the opportunity to determine their deficiencies in low level algebra skills and correct them before or during the course using the Internet. The program can discover that a student is having problems with the difference quotient because the student is not comfortable with complex fractions in arithmetic and then fix that deficiency. This benefits the individual student who is potentially more successful in the course, the instructor who can be more comfortable with students' current algebra skills, and the class as a whole because they can move through the material with more attention to ideas than to arithmetic and algebraic details.

In the future, such packages will include pre-calculus and calculus and will become an alternative to our current courses. We may also use them as a supplementary part of traditional courses to make sure that computational skills are maintained while we devote more attention to critical thinking skills and problem solving. Finally, using the assessment and tutorial features of such packages of instructional services, professors in other disciplines can be assured of the computational skills of students who have taken our mathematics courses as prerequisites to their courses.

An Internet Data Collection and Communication Device

As more and more applications of mathematics arise, the collection of data related to other fields for mathematical and statistical analysis will become more and more important. The current availability of sites where such data is available is growing at an enormous rate. The ability to use this information and for institutions to provide opportunities for students to participate in collaborative projects with students at other institutions is enticing. The availability of large scale computing facilities for students to access over the Internet will increase with the development of the Digital Libraries projects that are being funded by the NSF. The time for professors to avail themselves of these new opportunities seems to be decreasing, however.

Although the capabilities that the Internet provides and promises to provide are exciting, the human resources needed to make these accessible to students in an efficient, timely, and educationally appropriate manner are often difficult to orchestrate. On a smaller scale, the development of Internet courses that are for the first two years of collegiate mathematics but specialize in the topics of calculus for specific disciplines becomes possible.

At the moment many disciplines require two semesters of college level calculus (usually called calculus for science and engineering). These courses all seem to have applications, but not enough specific applications to satisfy any particular discipline. Biology, physics, mechanical engineering, electrical engineering, systems engineering, biochemistry, ecology, environmental science, decision science, economics, and psychology all have slightly different or vastly different requirements for what they need and want from calculus. The Internet provides an opportunity to tie such specialized laboratory courses to the traditional calculus courses taught at each school. Faculty from each of these different disciplines could then be satisfied that their students had seen the calculus in the particular light that they need while the mathematics faculty would not be forced to teach material that they felt uncomfortable with. For example, a consortium of colleges could agree that each college, following its strengths, would offer a laboratory course geared to one discipline and provide this course over the web. Students from each of the consortium schools could take the appropriate laboratory course for their discipline while taking a traditional course on campus. Mathematics students from each school could take several of these courses to fulfill the applied component of the mathematics major. In this way, the mathematics departments could freshen their offerings while providing mathematics majors with the applied background needed to obtain employment.

A Way of Performing Mathematical Computations with Numbers, Symbols, and Graphs

The second David Report of the National Research Council included many new breakthroughs in mathematics, a majority of which involved the use of computers. Although there are still mathematicians who do not believe that the four-color problem has been solved, much mathematics is done today using computers as computational tools for exploration, for completing and verifying complex numerical and symbolic computations, and for graphical presentation. Almost all major mathematical journals now accept papers in electronic form, some only in electronic form. Much of the mathematical literature is being converted to electronic form to improve its use and usefulness. The use of computers by mathematicians has become a commonplace activity, but such use in classrooms has not. That there was a time when mathematics was done in the sand and not with paper and pencil is not a surprise. That most mathematics in the next century will be done in silicon again would be a surprise to almost everyone in mathematics, but perhaps not to those in other disciplines who are already using the tools.

Conclusion

A challenge for us in the mathematics community charged with teaching the young is to create sets of activities that will provide our students with the understanding needed to use these powerful tools effectively to solve problems and develop theories that we have not envisioned ourselves. What and how much mental and paper and pencil activity do students need to understand the mathematics well enough to use a particular computer tool? Can we teach the mathematical concepts with the computer computational tools? What are the essential ideas that must be taught for the student to learn new ideas in a technology-rich environment?

The content of courses from calculus through linear algebra and differential equations could stand a complete reexamination based on the kinds of computational tools that are available to all our students in inexpensive, portable, and powerful devices. Some professors have begun to revamp the differential equations and linear algebra courses that can most benefit from extensive numerical computations. Calculus, the centerpiece of the undergraduate mathematics curriculum at many institutions, also could benefit from such review, in part because of the changes in the differential equations and linear algebra courses. The calculus reform movement has made many changes in the pedagogy of such courses and some changes have been made in the emphases of the course. In addition, most calculus textbooks these days have downplayed theory and rigor because of the perceived decrease in student ability. However, the notion that some topics should be de-emphasized or eliminated and the order of the topics rearranged has not been entertained.

In the future, a combination of the uses of technology suggested here will eventually be developed. Professors will create new courses with substantially different content, with in-depth projects from a variety of disciplines that rely on student computer computational capabilities that will be presented in classrooms (or not), and with video projectors to students that are Internet-connected. The student and the professor will rely on student mastery of prerequisites prior to the class meeting using ALEKS-like instructional services. How do we ensure that this new educational experience that will come to pass will be of high quality and provide the understanding and skill needed to further the mathematical enterprise?

Goals and Content Perspective

This section edited by Joseph Myers.

Content choices, balancing theory with computation, the diversity of the students in first-year courses, and the future role of calculus lead to fundamental questions concerning the intellectual goals of a mathematics curriculum. Over the past two decades, developing students to learn how to learn on their own has become accepted as central to the set of curricular goals. Although not identical in meaning, the phrases "life long learner", "learning to think", "mental discipline", and "learning the mathematical thought process" are used as synonyms for learning how to learn. The authors taking the Goals and Content perspective offer ideas that cover the spectrum from maintaining the status quo to replacing calculus with a program focusing on inquiry and modeling.

With respect to the choice of content, David Lomen and Paul Zorn express more satisfaction than do the editors and authors of Part 1 with the content in the present (reformed) calculus texts. They do not feel that content of current (reformed) calculus texts is the major problem with the calculus course. Jeffrey Froyd questions why very little of the mathematics developed in this century is found in core courses. He suggests using the question: "To what degree does topic X increase the capacity of a graduate to learn and create?" rather than the statement "Any graduate must know topic X." as the filter for determining content. The theoretical versus conceptual debate contrasts the pre-calculus reform (prior to 1985) thinking to the calculus reform thinking. Paul Zorn describes two poles in this debate as the *math way* —emphasizing limits as the major primitive and the *science way*—emphasizing rates of change as the major primitive. Jim Lightbourne provides a historical account of the calculus reform movement and parallels it with the present reform taking place in physics and engineering. He notes that the lack of communication and cooperation between departments restricts the effectiveness of the reform efforts in mathematics, physics, and engineering.

Student growth needs to be accounted for in curriculum planning, it is too important to be left to chance. Frank Giordano identifies learning how to learn, communications, mathematical sophistication, modeling, technology, connectivity, and history of mathematics as the important components of student growth. He offers a set of content objectives for a two-year integrated program that encourages progressive student growth in each of these categories.

The understanding and meaning of *high-standards* courses has changed from preparation for real analysis to ones that focus on deeper modeling experiences, open-ended projects, inquiry, and the ability to apply mathematics in interdisciplinary settings. Jeffrey Froyd suggests changing from the practice of insisting on a thorough understanding of prerequisite topics before introducing the next topic to a program that orders ideas around questions to be attacked. He states "The processes in which students participate can be as important as or more important than the ideas that are presented to the students."

Crossing the Discipline Boundaries to Improve Mathematics Education[1]

James Lightbourne[2]
National Science Foundation

Abstract. This paper indicates how changes occurring now in undergraduate science and engineering education can inform and support improvement in undergraduate mathematics education. Reports and discussions on education in sessions at national and professional society meetings have common themes and findings across the various disciplines. However, there is not much exchange of information across these discipline boundaries. Similarly, visiting college campuses, one frequently finds mathematics faculty who have more in common in terms of their views and practices to improve undergraduate education with faculty in physics, for example, than colleagues in the mathematics department. This lack of communication and collaboration across disciplinary boundaries results in missed opportunities that would benefit mathematics departments and mathematics education.

Section I provides a brief summary of the Tulane Conference recommendations and general trends found among the various calculus reform projects. Section II provides similar information from reports on undergraduate education in engineering and physics, to focus on the disciplines participating in this workshop. Section III describes projects in undergraduate physics and engineering education that illustrate specific efforts occurring in these disciplines. Section IV provides summary observations.

Material for this paper is drawn liberally from national reports and testimony to the National Science Foundation obtained during the *Shaping the Future* hearings [10, 11].

Calculus Reform

The "Tulane Conference" [1], with funding from the Sloan Foundation, was held in January 1986 in conjunction with the Annual Joint Mathematics Meetings. The conference, attended by 25 invited participants, identified five general problems encountered at that time in the teaching of calculus:

- too few students successfully completing calculus;
- students performing symbolic manipulations with little understanding or ability to use calculus in subsequent courses;

---

[1] This paper is part of an article by the author appearing in *Calculus Renewal: Issues for Undergraduate Mathematics in the Next Decade* (Susan L. Ganter, editor), Kluwer Academic Publishers (2000).

[2] James Lightbourne is Science Advisor in the NSF Division of Undergraduate Education. The opinions expressed in this paper are those of the author and are not intended to represent the policies or position of the National Science Foundation.

- faculty feeling frustrated with poorly prepared, poorly motivated students;
- calculus being required as filter through which other disciplines culled out students but made little use of calculus in their courses;
- mathematics lagging behind other disciplines in use of technology.

The mathematics community responded by developing new texts and other materials for teaching calculus. *Assessing Calculus Reform Efforts* [12] is a report on the findings of a Mathematical Association of America study to assess the calculus reform movement. Many of these calculus reform efforts have been supported by the National Science Foundation (NSF) through the NSF Calculus Program and other programs at NSF [4]. The materials offer a variety of approaches to teaching and learning calculus, reaching a broader student audience. Topics are presented through several representations; for example, graphical, numerical, symbolic, and written or verbal description. The changes in instructional practice include introduction — or increased use — of technology, modeling and applications, collaborative learning, student projects, student writing and oral presentations.

Reports from Other Disciplines

During this same period of time, other disciplines have been reconsidering how their subjects are taught. As the following examples illustrate, the concerns and recommendations for improvement parallel those in the mathematics community and, in particular, the calculus reform movement.

Engineering

NSF sponsored a workshop in June 1994, engaging 65 participants that represented engineering faculty, professional societies, industry, and students. The proceedings of that conference comprise the report *Restructuring Engineering Education: A Focus on Change* [8].

Concerns raised at the workshop included:

- classes typically taught in large lecture settings;
- problem assignments and assessments are highly structured;
- lack of research base in teaching and learning;
- lack of attention to different student career goals.

Proposed is creation of learning environments that include:

- active, collaborative learning;
- use of modules;
- research, development, and practical experience for undergraduates;
- learning-by-doing, the norm in professional fields;
- increased integration of science, mathematics, and engineering sub-discipline content;
- recognition of different backgrounds and career goals of students;
- rigorous educational research base in teaching and learning;
- appreciation for the complexities of physical devices and structures.

In terms of content, the workshop concluded that it is impossible to define an engineering curriculum applicable at all institutions. Each school needs to consider its own constituents and diversity of programs should be encouraged.

This flexibility in developing engineering programs is also reflected in the revised criteria used to accredit engineering programs by the Accreditation Board for Engineering and Technology, Inc. (ABET). Engineering Criteria 2000 [2] was approved by the ABET Board of Directors for a two-year comment period that began in January 1996. A phased-in implementation began with the 1998/1999 visit cycle, with full implementation of the Criteria 2000 in Fall 2001. The programs are evaluated based on student outcomes with specific course requirements not stated to the extent done so in past years. Evaluation evidence that may be used includes, for example, student portfolios, including design projects; nationally normed subject content examinations; alumni surveys that document professional accomplishments and career development activities; employer surveys; and placement data of graduates.

In the new criteria, engineering programs are expected to demonstrate that their graduates have the following capabilities:

- ability to apply knowledge of mathematics, science, and engineering;
- ability to design and conduct experiments, as well as to analyze and interpret data;
- ability to design a system, component, or process to meet desired needs;
- ability to function on multi-disciplinary teams;
- ability to identify, formulate, and solve engineering problems;
- understanding of professional and ethical responsibility;
- ability to communicate effectively;
- broad education necessary to understand the impact of engineering solutions in a global/societal context;
- recognition of the need for and an ability to engage in life-long learning;
- knowledge of contemporary issues;
- ability to use the techniques, skills, and modern engineering tools necessary for engineering practice.

The changes in the ABET evaluation criteria - and the consequent changes in engineering education — potentially could have a significant impact on mathematics departments. As indicated earlier, the new criteria are based on expected student outcomes, rather than on a checklist of courses. This presents an opportunity for mathematics departments to work with the engineering departments toward the outcomes-based criteria. However, the new criteria also present the danger of losing the teaching of courses for those departments that do not do so.

Physics

There is a growing group of individuals conducting discipline-based research in teaching and learning in physics. The paper *Resource Letter on Physics Education Research* [6] provides an annotated compilation of over 200 references, primarily focused on postsecondary education. Many of the references are empirical studies that consider student understanding of a specific topic. *Teaching Physics: Figuring Out What Works* [7] is an example of a more general paper. This paper poses three questions: what is involved in understanding physics? what do students bring to physics classes? how do students respond to what they are taught? Among other findings, the paper reports results from a study comparing three educational environments: traditional (large lectures with small group recitations and laboratories), tutorials (including student group work on research based worksheets), and Workshop Physics (lectures, recitations, and laboratories combined into lab-based sessions)

A large body of research involves use of a multiple-choice diagnostic test, the Force Concept Inventory (FCI) [5]. FCI is a 29-question test that assesses student understanding of concepts in mechanics. Studies [e.g., 7] reported in the literature show that student performance on FCI does not necessarily improve with traditionally taught classes; in fact, student performance actually appears to have deteriorated. In addition, students appeared to deteriorate with traditional instruction in general areas such as learning independently, linking physics with reality and mathematics, and understanding concepts. An extensive study [3] conducted in a variety of school settings concluded that students in interactive classes consistently scored better on diagnostic tests than students in traditionally taught classes.

Testimony given during the hearings for *Shaping the Future* [13] reported that "we know beyond any reasonable doubt" that engaging undergraduate students in active learning and active research, in close contact with faculty and other students, encourages students of all kinds to continue toward a career in science. Students are engaged in active learning through several means:

- classroom instruction that keeps students active;
- early participation in research;
- appropriate use of technology – for example, interconnected computers provide focus for small group discussions; spreadsheets provide means for numerical computation; digital video processing provides means to study realistic applications.

Concern was expressed that 1) computer simulations of experiments easily conducted in laboratory and 2) computer-aided instruction as was traditionally implemented, isolate students and do not have desired outcomes. Also, the demand for coverage of material too often outweighs the demand for conceptual understanding and true learning.

Examples From Other Disciplines

The following are a few projects that have recently been funded by NSF in engineering and physics that illustrate in more specific terms some of the changes occurring in other disciplines.

- The University of Florida is developing a real-time interactive flight test program. The program allows participants to perform airborne experiments, with data reduction and analysis in real-time on board the aircraft. Videos and other sensor data from the aircraft are sent to classrooms via video-conferencing, so that the classes may actively participate in a real-time flight test.
- Daytona Beach Community College is creating a new instructional environment for introductory courses in electronics, computer-aided design, civil engineering, and computer programming. The objective of the project is to develop a virtual classroom environment through which students can access course materials and interact with other students and faculty.
- The conception, production, evaluation, and dissemination of a series of interactive modules for the teaching and learning of fluid mechanics in science and engineering is being developed jointly by Stanford University, Massachusetts Institute of Technology, and the University of Illinois. The modules focus on fundamentals of design but could be used in curricula of other disciplines. The primary objectives are to enhance student problem solving, intuition about complex flow phenomena, and retention of knowledge.
- A microcomputer-based laboratory at the University of Maine, Farmington is being used to introduce an inquiry-based curriculum into the general physics sequence. The student population of the course is approximately half science majors and half secondary education majors. In this project there is a particular focus on the pre-service teachers in the course. This focus consists of having "alumni" of the general physics sequence return as peer instructors in both the workshop physics course and the conceptual physics course designed for non-majors. This is being done by having the physics and secondary education faculty work together to affect program changes that would require the secondary education students to have this teaching experience as part of the degree requirements in education. The goals are to further improve the understanding of physics of these science teachers-to-be and to give them some practical experience with an inquiry-based physics curriculum.
- A project at Carnegie-Mellon University is developing a calculus-based introductory textbook on mechanics and thermal physics. The text presents ideas previously treated separately as an integrated whole, emphasizing atomic-level description, analysis, and modeling.

Summary Observations

The previous sections serve to illustrate common concerns and recommendations for undergraduate education engineering, mathematics, and physics. In summary, the following major areas are identified:

- *Course Emphasis.* Current courses tend to emphasize manipulative skills, routine experiments, or cookbook techniques rather than student understanding and competence in the subject. Course design, including student testing, should place more emphasis on understanding of concepts, scientific method, and relevance in a broader context. The curriculum in general should reach a wider spectrum of students in terms of backgrounds, interests, and career goals.
- *Educational Practices.* In general, there is too much reliance on lecture, routine student exercises and laboratories, and examinations designed primarily to minimize student and faculty time. The reports recommend that faculty increase their use of collaborative learning, discovery based student activities and student research, projects, writing assignments, oral presentations, and other practices that provide more engaging and effective education.
- *Computer Technology.* Computer technology can be used, for example, to engage students in discovery, provide access to large databases, gather information, and collaborate over large distances.
- *Content.* In general, these reports conclude that courses try to cover too much material, at the expense of a sufficiently deep treatment of the subject. It is recommended that courses include fewer topics for which deeper student understanding would be possible and expected.
- *Research Based Educational Decisions.* Discipline based research in teaching and learning should provide a scholarly basis for informed educational decisions. This research is providing insights into what students actually learn and what educational practices are effective for improved learning.

At disciplinary society meetings across the country, faculty discuss ways to improve undergraduate education. Although these meetings are held within the separate disciplines, the issues, concerns, and recommendations have much common ground across the disciplines. These faculty return to their home institutions and too often find themselves working in isolation. Colleagues in their own discipline may not be receptive, and they do not communicate across the discipline and department boundaries.

The benefits of these interactions across discipline lines are multiple. Students benefit in having both content and pedagogical approaches, including uses of technology, reinforced in different courses and discipline settings. The content in mathematics courses can be enriched through applications relevant to the other courses that students take. Faculty implementing similar strategies in different disciplines can benefit through collaboration. The institutional support possible through having a critical mass of faculty with common interests is also not realized. Moreover, the mathematics department, in general, can be better positioned as central to the institutional mission to provide undergraduate education.

References

1. R. G. Douglas(ed.) 1986. *Toward a Lean and Lively Calculus, MAA Notes No. 6*, Mathematical Association of America.
2. *Engineering Criteria 2000*, Engineering Accreditation Commission of the Accreditation Board for Engineering and Technology, 1995.
3. R. R. Hake, Interactive-engagement versus traditional methods: A six-thousand student survey of mechanics test data for introductory physics courses, American Journal of Physics **66,** 64–74, 1998.
4. W. E. Haver (ed.), 1998. *Calculus: Catalyzing a National Community for Reform,* Mathematical Association of America.
5. D. Hestenes, M. Wells, and G. Swackhammer, Force Concept Inventory, Physics Teacher **30,** 141–158, 1992.
6. McDermott, Lillian C. and Edward F. Redish, Resource Letter on Physics Education Research, preprint.
7. Redish, Edward F. and Richard N. Steinberg, Teaching Physics: Figuring Out What Works, Physics Today **52**, 24–30, 1999.

8. *Restructuring Engineering Education: A Focus on Change,* National Science Foundation, NSF 95–65, 1995.
9. *Science Teaching Reconsidered: A Handbook*, National Academy of Sciences, 1997.
10. *Shaping the Future: New Expectations for Undergraduate Education in Science, Mathematics, Engineering, and Technology*, National Science Foundation, NSF 96-139, 1996.
11. *Shaping the Future, Volume II: Perspectives* for *Undergraduate Education in Science, Mathematics, Engineering, and Technology*, National Science Foundation, NSF 98-128, 1998.
12. A. C. Tucker and J. R. C. Leitzel 1995. *Assessing Calculus Reform Efforts,* Mathematical Association of America.

Thoughts on the Next Challenge Regarding Calculus Goals and Content

David O. Lomen
University of Arizona

Abstract. This paper takes the view that goals, content, and pedagogy are inseparable. We see that the role calculus took on as a filter was not only due to inadequate classroom materials (mostly unmotivated and pedantic) but also in response to student problems with mastery of prerequisite skills, lack of quality study time, and lack of appreciation for the importance of calculus in their downstream courses. Despite vigorous reform efforts, these problems are still with us. In this paper we suggest some ways of attacking these outstanding fundamental problems and then consider how the content and pedagogy will finally be able to successfully evolve in response. If we remain unable to address these fundamental problems, "calculus as filter rather than pump" may still be true in the next decade as well.

Calculus content cannot be separated from pedagogy and strongly depends on what goals have been established. These goals should be determined in cooperation with instructors in the disciplines populating these courses. Some typical goals are to:

- Develop competent, confident, creative, problem solvers who can use calculus to model situations outside of mathematics;
- Develop the analytical skills and reasoning ability which meet downstream requirements for courses using calculus as a prerequisite (both in mathematics and partner disciplines);
- Increase students' independence and confidence which will help develop habits needed for their lifelong learning.

In order to help students achieve these goals, the materials used should:

- Be application driven (rather than axiom driven);
- Have a logical development of ideas — including critical thinking, nature of evidence, and mathematical reasoning;
- Emphasize the ability to
  - i) check reasonableness of answers,
  - ii) use proper technology,
  - iii) generalize and transfer ideas and techniques to new situations.

Of course the requirements of the department and partner disciplines regarding the use of technology, and its availability to students, will also influence the content of the calculus courses.

While use of technology and nontraditional materials have addressed some of the problems considered as the basis of the dissatisfaction expressed at the "Calculus for a New Century" conference in 1987, very few, if any, educators today think all of those problems have disappeared. Three problems which still seem to be present at many institutions are:

- A lack of mastery of prerequisite skills for the specific course.
- A lack of student willingness to spend sufficient quality time studying calculus.
- A lack of awareness as to the level of calculus proficiency needed in subsequent mathematics, science, or engineering courses.

To counter this student-centered problem, along with the challenge of how to best use the increasing availability of access to the World Wide Web is the current task of calculus committees.Four ways of addressing the lack of prerequisite skills for first semester calculus are to:

1. Increase the number of meeting times.
2. Teach a combined pre-calculus/calculus course.
3. Administer a "Readiness" exam where failure to pass disbars the student from the course until they have mastered the prerequisite material.
4. Provide a 2-unit review of algebra/trigonometry class to which students can step back after doing poorly in Calculus I for the first four weeks.
5. Require students to review this material outside of class. Current software programs or material on the web may ease instructor responsibility for this.

The lack of prerequisite skills is not just limited to first semester calculus. A major problem for some students is that they do not take Calculus I, Calculus II, and Calculus III at the same institution, or if they do, do not take them in consecutive terms. The half-life of much of what students learn in calculus is thought to be on the order of months, not years. Having gateway exams for all calculus courses is one way of addressing that problem. (Several faculty at the University of Arizona have used such exams—based on the "Are You Ready? Software" [10]—and obtained substantially lower failure rates. A few of the students who did poorly on this exam dropped back to the prerequisite course, but the majority of them became serious, studied, worked on regaining lost skills, and were successful in the course.) Several schools already have such exams in place. The University of Nebraska's website [18] is one place to check. Another remedy is to hold extra review sessions over prerequisite material, or have study sessions run by undergraduate or graduate students. The current round of VIGRE grants provides money for Undergraduate Teaching Assistants. One way to use these undergraduates is to have them lead these problem, review, or discussion sessions, thereby freeing up more time in class to discuss concepts or in-depth applications. They can also be used to facilitate group work within the classroom or with assignments designed to be completed in a computer laboratory.

If students realized the extent to which they would use their calculus knowledge in subsequent courses they would spend more time studying calculus. Here is where a good textbook can help, if it includes challenging exercises from partner disciplines. Including major projects in the homework such as those found in [1], [5], and [8] is what many institutions have done to address this problem. In the future, we may have some help here from our partner disciplines. Many engineering faculties are stressing that because of the rapid changes in their field, their students need to learn how to learn. The usual label of such endeavors is to learn to become a lifelong learner. We also need to teach our calculus students how to learn. Having a major focus in calculus on understanding concepts is in keeping with this goal.

If the above issues are not addressed, the observation that calculus is a filter, not a pump may still be valid ten years from now. If these issues are addressed, we may see the following happen.

1. We will spend less time in the classroom on routine exercises and more time on developing concepts and ideas. Tutorials on standard material will be available from the web to help facilitate this (see the website of Harvey Mudd College [19] for examples). With an increase in the use of visualization made possible by technology (including the web) we can enhance students mathematical maturity so they will not consider calculus to be a collection of black box operations. Rigorous thinking [11] will become more important, but this is not to be construed as having students regurgitate standard proofs. Instead we will use simulations or graphical and numerical examples to show the need for theory, and then use a logical development to show why the needed result is true. This may or may not lead to a rigorous proof but the students should be convinced of the result. (As an aside, it appears that using technology to enhance conceptual understanding is one measurable improvement this decade [7].)

2. Increasing use of Computer Algebra Systems (CAS) will continue to keep the time spent on techniques of integration to the current levels of most reform textbooks. However, the need for basic substitution, integration by parts, and partial fractions will continue, because they are needed for the current way many courses in mathematics, science, and engineering are being taught. They will always be useful in theoretical developments in these later courses as well. We will also need to construct examples that encourage careful reasoning, with or without CAS. The appendix contains such an example, which students find interesting, that incorporates use of the chain rule with either the first or second derivative test for maxima and minima.

3. The value of knowing the accuracy of numerical techniques of integration will also continue to be emphasized. Taylor polynomials will have added importance, but other than the ratio test and comparison test, little will be needed at this level in the area of convergence of infinite series.

4. More material from sources other than the textbook will be used. I went through the past four years of Primus and include in my list of references many of the novel ways of introducing theoretical and practical matters to calculus students. There are also some projects here, as well as the extensive source of projects mentioned earlier. Modeling and applications will be crucial to helping achieve the goals mentioned in the first paragraph. The web will be a place for motivational examples and simulations, with "just in time" links to appropriate background material, techniques or theory. This is the exciting new frontier in education, and is the topic of other articles in this volume.

In conclusion I note that more and more students take calculus in high school, and often enter our calculus courses with better backgrounds than our continuing students. Here we have an opportunity to utilize their background by offering special courses. For example, for those entering students who score a 4 or 5 on the AB exam, we have, for the past four years, offered a two-semester course which starts with simple differential equations and proceeds to the stability of autonomous systems before tackling improper integrals and other topics from Calculus II. After covering all of Calculus II (motivating Taylor polynomials and infinite series by examining differential equations which do not have "nice" solutions), the course concludes with more topics from differential equations. We feel the success of this course is due to the order of topics, which has the advantage of having students start the semester with new, interesting, and challenging material, instead of re-hashing integration techniques and applications at the beginning of the standard Calculus II course. For those students who score a 4 or 5 on the BC exam, maybe we could have a combined Vector Calculus/Linear Algebra course like we had about thirty years ago. We tried this, but it was not a success because of the lack of a textbook that students were motivated to read. A book is needed that follows the lead of some of the current reform texts in calculus and linear algebra.

References

1. D. C. Arney (Editor) 1997. *Interdisciplinary Lively Application Projects(ILAPS)*, MAA.
2. G. Ashline and J. Ellis-Monaghan. "Interdisciplinary Population Projects in a First Semester Calculus Course", *PRIMUS*, Volume IX, No. 1, March 1999 39–55.
3. J. Beidleman, D. Jones, P. Wells. "Increasing Students' Conceptual Understanding of First Semester Calculus Through Writing", *PRIMUS*, Volume V, No. 4, December 1995, 297-316.
4. M. Branton and M. Hale. "Interactive Texts and a Virtual Environment for Exploring Spring-mass Systems", *PRIMUS*, Volume VI, No. 1, March 1996, 61–67.

5. M. Cohen, E. Gaughan, A. Knoebel, D. Kurtz, D. Pengelley, 1991. *Student Research Projects in Calculus*, MAA.
6. D. Ensley. "Capping the Calculus", *PRIMUS*, Volume VI, No. 3, September 1996, 269–276.
7. S. Gantor 1999. *Ten Years of Calculus Reform: A report on evaluation efforts and national impact, in Assessment Practices in Mathematics*, B. Gold (editor), MAA Notes, Volume 49.
8. S. Hilbert. J. Maceli, E. Robinson, D. Schwartz, S. Seltzer 1994. *Calculus: An Active Approach with Projects*, John Wiley and Sons Inc.
9. A. Klebanoff. "A Memorable Drive Through Calculus", *PRIMUS*, Volume VII, No. 4, December 1997, 289–296.
10. D. Lomen and D. Lovelock. "Are You Ready?" Disks: An Aid to Precalculus Reform, in *Preparing for a New Calculus,* MAA Notes 36, Ed. A. Solow, 1994, 145–148 (available from `http://www.math.arizona.edu/software/uasft.html`).
11. W. McCallum. "Rigor in the Undergraduate Calculus Curriculum", *AMS Notices*, Volume 38, No. 9, 1991, 1131–1132.
12. D. Pirich. "A New Look at the Classic Box Problem", *PRIMUS*, Volume VI, No.1, March 1996, 35–48.
13. M. Revak, D. Pendergraft, and C. Brown. "The Great Calculus II Conundrum", *PRIMUS*, Volume VII, No. 3, September 1997, 243–348.
14. S. Sprows. "What Can Happen When the Second Derivative Vanishes", *PRIMUS*, Volume VI, No. 4, December 1996, 381–384.
15. K. Weld. "Splines and Roller Coasters: A Calculus Project Using MAPLE", *PRIMUS*, Volume VI, No. 2, June 1996, 97–106.
16. A. Young. "Two Familiar Theorems: Modified So Their Proofs Are Comprehensive To First-Semester Calculus Students", *PRIMUS*, Volume VI, No. 2, June 1996, 107–116.
17. A. Young. "Discovering the Derivative of an Exponential Function: A Calculus I Project", *PRIMUS*, Volume VII, No. 1, March 1997, 18–24.
18. `http://calculus.unl.edu/gateway.html`
19. `http:// www.math.hmc.edu/calculus`

Millennial Calculus Courses: Goals and Content

Paul Zorn
St. Olaf College

Abstract. In this paper, we consider the balances between several possible pairs of paths of emphasis for our core programs: between concept and application, between a math point of view and a science point of view, and between skills and concepts. In each area we tend toward the eclectic, with a bias toward a visceral understanding of concepts over a mathematical firmness of reason. This leads us to propose a content sketch for a modern, beginning math program. Finally, we interpret the community's hunger for high standards within the context of the utility of our discipline to our students, and argue for more rigorous nontraditional activities to fill this role.

Goals and Content

What are math/science/engineering-oriented calculus courses for? What should calculus students know? What should they do? What applications should they see, and why? How will we help students achieve our goals for them? How are these questions related to each other, and to the present and future state of mathematical computing? What happens next?

The goals and content of calculus courses are, of course, tightly intertwined. To a large extent, the content of a course represents the instructor's practical strategy for achieving its larger goals. So I'll discuss goals and content mainly together, mentioning either or both as they arise.

Some legitimate goals of calculus, however, have little to do with specific content choices. For instance, mathematics courses in general, and calculus courses, in particular, are sometimes said to build mental discipline, introduce the mathematical method, and (more grandiosely) teach students how to think.

This sort of talk sounds a little passé nowadays; we're probably more comfortable talking about interesting applications of calculus and about how advanced mathematics builds fundamentally on calculus. But the old "mental training" agenda is still perfectly valid, in my opinion. It may gain even more weight in the future, as both applications and the field of mathematics change, perhaps in ways we haven't yet dreamed of. Mathematical computing is the obvious change engine right now, but it's far from obvious that that will continue. What's ahead, for example, if genetics and neurobiology continue to grow at their present rates? Today's applications may soon seem quaint, but mathematical ways of thinking will endure.

Calculus for Whom?

In prognosticating about what calculus courses should do and be, we should be clear about our audience(s). Are we addressing mathematics majors? Engineering majors? Students who've already seen some calculus in high school? Well-designed calculus courses need to take due account of the different needs of these different groups.

One possibility is to choose a specific clientele, focusing narrowly on its special needs and interests. Another is to address a diverse clientele (as is usual in liberal arts colleges), stressing broad ideas and principles that cut across areas of application. Which of these situations obtains governs many important design choices: whether mathematical models are "built" or only "used"; the level and role of proofs and mathematical rigor; the importance of limits; the choice of applications; the balance of numerical and symbolic approaches; the curricular prominence of DEs; the importance of developing symbolic facility (by hand, by head, or by machine); etc.

My hunch is that, with some important and useful exceptions (including the very sort of integrated mathematics/science courses being discussed right here at this conference), most beginning calculus courses will continue to serve relatively broad student audiences, and will have relatively broad educational goals. Calculus can be many things: a general tool for mathematical applications; the end of a long and tedious road through pre-calculus; an entry point to the mathematics major; a "general education" introduction to mathematical culture and thought. For many students, the course serves several of these functions.

I would hazard the companion guess, therefore, that mainstream calculus courses of the future will hew more, not less, to basic mathematical ideas and concepts (which serve a wide and growing range of disciplinary and pedagogical goals) than to specific areas of application. If I'm right, a challenge for the future will be to find not only new and attractive applications of calculus but also, and just as important, ways of making the good, old, useful, basic ideas of calculus clearer and more applicable.

The Two-Fold Way

In thinking about calculus content and goals, it may be helpful to recall two quite different basic approaches to the subject. One strategy—let's call it the *math way*—is to emphasize limits, which are indisputably the fundamental mathematical objects on which a rigorous development is based. Another road to the calculus—let's call it the *science way*—is to take rates of change, not limits, as the course's main "primitives," leaving to later work (and, therefore, mainly to mathematics majors) such subtler and less intuitive issues as defining rates rigorously (as limits!), proving existence, and the like.

To illustrate the difference, consider the tangent line problem. The science way—zoom in on a nice graph until it looks straight—takes existence questions for granted (or defers them to later courses), and worries mainly about meaning and interpretation. The math way forces the hard questions about definition and existence, and answers them using limits.

The science way has clearly been in the lead for the last 15 years, since what we call calculus reform came along. It may be less obvious, however, that the science way has been on the rise for over a century, with calculus authors (including such true luminaries as Augustus de Morgan) arguing for a physically intuitive, rates-based approach, rather than one based on a rigorous theory of limits. The same rates-based approach characterizes Elias Loomis's *Elements of Analytical Geometry and of the Differential and Integral Calculus*, one of the first American textbook treatments, published in 1852.

I like the rates-based approach to beginning calculus, and I see no sign of it changing. (This may sound like heresy, coming from a mathematician, and an analyst to boot. But I think that mathematical analysis, like certain other rare and keen pleasures of life, is best enjoyed at an appropriately advanced age.) One of our tasks for the future may be to find problems and applications that both build on and advance a rates-based understanding of the subject.

Symbolic Facility and Symbol Sense

What balance of skills and concepts is necessary to use calculus effectively in science and mathematics? I find the question highly non-trivial—especially in the presence of symbol-manipulating technology tools. One answer—that once a machine can do something, humans shouldn't—I find neither pedagogically nor practically convincing. (In a vaguely related vein, I've heard that the U.S. Naval Academy still requires beginners to master sail-driven craft—a nice idea.) The key, I think, is to to *link* skills with concepts, choosing the former to strengthen the latter.

For instance, I see little value in spending one's limited course budget of time and energy on complicated partial fractions problems (e.g., ones that involve powers of irreducible quadratics)—that may as well be left to *Maple* or *Mathematica*. But the idea of partial fractions and some simple examples are probably as important as ever. There's little typographical difference between the functions

$$f(x) = \frac{1}{1-x^2} \quad \text{and} \quad g(x) = \frac{1}{1+x^2}$$

but, in every other way, the functions (and the phenomena they might model) are completely different, as are their antiderivatives: the functions

$$\int f(x)dx = \frac{\ln|1+x| + \ln|1-x|}{2} \quad \text{and} \quad \int g(x)dx = \arctan(x) .$$

The partial fraction decomposition goes a long way toward explaining what's happening. (Looking at graphs is helpful, too, of course, but graphs alone don't explain why logs and arctangent, rather than other functions with approximately the same shapes, are involved.)

For some students it's enough, at least on the first pass, to learn mainly *about* calculus, concentrating on its main objects and ideas more than on its manipulations. Most mathematics and science students, on the other hand, have a more ambitious facility agenda—they need, sooner or later, to develop enough speed and confidence to do calculus efficiently, and use it as a sharp tool to solve problems in other areas.

This sort of mathematical facility has always required some skill and ease in throwing symbols around on paper or in one's head. I think it always will—even when every student has symbol-manipulating technology at her elbow. (Or perhaps on her wrist ... a small company in my hometown is busy developing an Internet-ready Dick Tracy watch.)

What symbolic facility means may well change. In the past, it meant things like expanding complicated rational functions in partial fractions, factoring polynomials, or grinding out tricky symbolic antiderivatives by hand. In the future, symbolic facility may mean other things, such as anticipating the form of an answer (e.g., the logarithmic and arctangential ingredients in rational function antiderivatives, the nested form of a composite derivative, etc.), or noting its absence with concern.

To help students build this facility, or symbol sense, we will probably continue to assign symbolic exercises, but perhaps in new forms, designed to foster pattern recognition and to point out structure as much as to crank out symbolic results. For example, I'd like to see students understand and visualize better the effect of parameters on function families like functions $f_a(x) = ax + \sin(x)$ and their derivatives. I'd gladly trade some of this for, say, trigonometric substitutions.

We might also aim to connect symbolic representations and operations more directly and concretely to graphical representations—a goal made more attainable by technology that handles numbers, pictures, and symbols.

Content

What specific content changes and emphases might support a modern beginning calculus course for science and mathematics? Here are some guesses:

Differential equations. Although differential equations and initial value problems are unquestionably the most useful calculus tools for the clientele at hand, they've played oddly minor roles in calculus courses so far. The basic ideas behind DEs, IVPs and their solutions are entirely accessible to calculus students, even if sophisticated symbolic solution techniques are not. With graphical and numerical tools available, DEs will probably achieve the more prominent place they deserve.

Infinite series. Modern beginning courses may well de-emphasize infinite series, at least as regards abstract questions of convergence and divergence. At the same time, more attention may be due to (related) issues of approximation. For example, whether a Taylor series approximation actually converges to the right place may be less of interest than how closely Taylor polynomials approximate the target function.

This is by no means to deny that series, convergence, and divergence are lovely "proto-analysis" topics for mathematics majors. But for most students these topics belong at a later developmental time.

Multivariate topics. We may soon pay more and earlier attention to multivariate topics of various sorts. These could include multiple integrals, partial derivatives, parametric equations, and the like. These are probably more important to most non-mathematics majors, and more likely to be encountered later, than (say) abstract convergence and divergence, which they might well displace in one-year courses.

Discrete and dynamical systems. Difference equations are natural counterparts to DEs. With even modest computing resources at hand, students can investigate discrete dynamical systems, use them in modeling, and perhaps begin to see how discrete and continuous viewpoints complement each other.

Setting High Standards

Among the bum raps sometimes adduced against calculus reform, the "low standards" rap may be the bummest. Ask almost any student, almost anywhere, who has experience in both standard and reformed courses, which is *harder*. Still, the question of high standards is fair. What are they? How will we achieve them?

High standards are, at one level, like motherhood and apple pie—who could oppose them? But what does the phrase mean? What *should* it mean? In times past, high standards in calculus (e.g., in honors calculus) tended to mean traditional mathematical rigor: precise definitions, careful proofs, development based carefully on limits, and mainly mathematical applications. High-standards calculus was, in essence, the fast track to real analysis.

But there's a broader view of high standards, toward which I think we're (appropriately) tending, driven partly by computing possibilities and partly by the different uses our students will make of calculus. High-standards calculus courses of the near future could be characterized by such things as more and deeper modeling problems (where students build, not just use, calculus-based models); more writing and verbal presentation; more open-ended, investigative activities; more challenging applications. (The traditional proto-analysis course will continue to exist, but as only one option.)

High standards are not only for the gifted—they should apply in all courses. The difficult material in standard courses has traditionally been symbol manipulation—trigonometric substitution, complicated partial fractions, nested composite derivatives, some timid stabs at epsilon and delta. But those aren't the only possibilities, or the best ones. We can challenge all students, not just in honors courses, with appropriate versions of the features mentioned above. All students need to understand what they're doing and why—not just how to calculate.

First Two Years of Mathematics for Scientists and Engineers

Jeffrey E. Froyd
Texas A&M University

Abstract. This paper challenges our discipline's long held but unspoken assumption that competence in mathematical skills and topics paves the road to success for our graduates who enter professions that use mathematics. This is misleading; rather, we argue that it is the ability of our graduates to learn and create that becomes their most crucial competency. We claim that "what math skills and concepts should our graduates learn?" is the wrong question, and reframe it more properly as "what math topics and activities will best increase the capacity of our graduates to learn and create?" This leads us to propose that more appropriate content for our courses synthesizes concepts from several disciplines, and is geared more toward desired student thought processes rather than topics.

Consider the following proposition "In the long run, your only sustainable source of competitive advantage is your ability to learn and create faster than others." The proposition is the individual analog to the proposition that Peter Senge advocates for organizations, "In the long run, the only sustainable source of competitive advantage is your organization's ability to learn faster than its competition." The rapidly growing amount of information also supports the proposition. People have speculated that the half-life of the knowledge that a scientific or engineering graduate gains during a bachelor's degree is 2–5 years. Then, after ten years at least 75% of the factual database that the graduate gained in school is obsolete and must be augmented by new knowledge gains in the workplace. Although the correctness of the proposition could be debated at length, let's accept the proposition as true for present purposes.

If true, then it is not what the graduate knows at graduation that is most important, but the rate at which the graduate can learn and create after graduation. Therefore, in discussions about content and whether to include topic X, the criterion for inclusion should not be "Any graduate must know topic X." or "Topic X is fundamental to the study of mathematics, physics, chemistry, engineering, etc." Instead, the criterion that follows from the proposition is the following, "To what degree does topic X increase the capacity of a graduate to learn and create?" With temporal constraints imposed by a four-year degree, the criterion could be restated, "What limited set of topics provides the largest increase in the capacity of a graduate to learn and create in his/her chosen field of study?" The difference between the two types of criteria is crucial because the list of topics that satisfy the criterion "Topic X is fundamental to the study of mathematics, physics, chemistry, engineering, etc." is very long and growing. Using the "Topic X is fundamental" criterion encourages long, highly charged debates because every faculty member has her/his set of topics that are fundamental. Since the union of the sets of topics from three or four diverse faculty members probably exceeds the limits of a four-year curriculum, debates on topic choice using "Topic X is fundamental" will never conclude. A different criterion is needed and a criterion based on the capacity of a graduate to learn and create is proposed.

What are mathematical topics that increase the capacity of graduates to learn and create? It is suggested that the topics that most increase the capacity of graduates to learn and create are related to the abilities to

- envision desired behavior and physical implementations that realize the desired behavior,
- abstract behavior that they observe in the world around them,
- formulate these abstractions carefully and quantitatively,
- predict behavior from the quantitative models, and
- apply the knowledge of the predicted behavior to improve their understanding of and their ability to predict behavior in the world around them.

Topics related to the ability to start with one symbolic formulation and derive symbolic formulation do not seem to increase the capacity of graduates as much as topics related to the previously listed abilities. Are topics such as the ability to derive symbolic indefinite integrals important? Absolutely, but importance is not the agreed upon criterion for inclusion. Instead, the criterion is the ability to increase capacity to learn and create.

Let's now explore the consequences of using the criterion to build the content of a mathematics curriculum. Does the knowledge of the concept of a function increase the capacity to learn and create? It is suggested that the answer is only slightly. Knowing the definition of a function may allow students to pass a test. However, knowing the definition does not appear to increase the capacity to learn and create, as much as the ability to abstract observed behavior or measured data by constructing functions to capture the essence of the behavior or the data. For example, can students construct functions that model a quantity to be optimized? Can students create functions that describe motion or other observed behavior? Does knowledge of symbolically constructing derivatives of expressions increase the capacity to learn and create? Again, it is suggested that the answer is only slightly. This is especially true since we now have software that performs symbolic manipulation faster and more accurately than humans. Instead, students that can apply the derivative to create quantitative abstractions or models, e.g., differential equations, have the greater capacity to learn and create. Similar observations can be made about computing limits of expressions, sequences, and series; symbolically computing integrals; or solving differential equations. What graduates need to increase their capacity to learn and create is the ability to express questions mathematically, reason with the created abstractions, and translate mathematical results into powerful ideas. In other words, students must use mathematics as a second language to describe, understand, and predict behavior in the world around them.

Similar content questions can be explored in other areas of mathematics. Is it more important to express the idea of getting close to something or to express relationships between rates of change in a system? Is it more important to express the idea of finding closed-form expressions for integrals of symbolic expressions or to express the idea of adding acceleration to get velocity, velocity to displacement, force dotted with displacement to get work, electric field dotted with displacement to get potential, [magnetic field], etc? Is it more important to express the idea of determining whether and where a sequence of numbers ends or to express the idea of determining quantitative models for a set of data? Is it more important to determine area under a curve or create a sum that estimates the weight of a column of air in the earth's atmosphere, or a sum that estimates work done along a curve, or a sum that estimates gravitational force due to a region or matter? In the spirit of an interdisciplinary approach, it is suggested that communities of mathematicians, scientists, and engineers should formulate answers to these and other questions. Once a list of important ideas to express mathematically have been developed, then the next task is to order the list in which students will learn to express these ideas and finally to develop activities through which students can gain expressive fluency.

Let's consider another example in which we can apply our criterion in selecting and organizing the content of a mathematics curriculum. After graduation, people frequently encounter sets of data. Data may be offered to support a proposition or data may be measured behavior of a system. People then must either interpret the data to determine the degree of support that it offers for a proposition or use the data to construct a model for the system. It is suggested that the capacity to learn and create will be significantly increased by the ability to interpret sets of data or use them to build models. Methodologies that support interpreting data or empirical modeling fall within probability and statistics. Therefore, calls are made to include a course in probability and/or statistics. However, faculty who teach these courses point out that students who enter the course have little or no previous exposure to concepts of randomness and risk. Further, they describe the difficulty in changing the deterministic orientation of students in a single course. A traditional answer is multiple courses, but this answer encounters many obstacles in already overloaded curricula. An alternative approach would be to introduce randomness and analysis of data early in a mathematics curriculum and spirally build concepts over the course of the curriculum. The integrated approach appears to answer the difficulty in helping students construct a stochastic framework in a short period of time and providing them increased capacity to learn and create.

Once we have agreed that the criterion for including a topic in our two-year mathematics curriculum is increasing capacity for a graduate to learn and create, is there a compelling reason to limit the possible topics for inclusion to calculus and differential equations? Isn't it possible that topics in linear algebra, discrete mathematics or statistics will give graduates more capacity to learn and create than simply including more topics from calculus and differential equations? Physicists, who have worked on a related approach to the introductory, one-year physics curriculum, have developed a curriculum called "Six Ideas that Shapted Physics." These ideas are

- Conservation Laws Constrain Interactions
- The Laws of Physics are Universal
- The Laws of Physics are Frame-Independent
- Electromagnetic Fields are Dynamic
- Matter Behaves Like Waves
- Some Processes are Irreversible

Choice of these topics met the guiding principles of the committee that developed this approach: a) less is more, b) include twentieth-century physics, c) use a story line, d) pay attention to the results of educational research, and e) seek the middle way. Although not every topic from previous physics curricula may fit under these six ideas, physicists who developed this curriculum thought that fluency with these six ideas would increase the capacity of graduates to learn and create in the future. Following the lead of these physicists, what are the six ideas that shaped mathematics?

Gary Sherman, Department of Mathematics, Rose-Hulman Institute of Technology, describes four powerful ideas that shape his view of mathematics: 1) measurement, 2) measurement in the presence of structure, 3) equivalence, and 4) transformation (go someplace, do something, return). He presents an informal explanation of the four ideas as they relate to discrete mathematics.

> Counting (measurement) in a raw set (of real-world size) is a mind-numbing, indeed computer-numbing, chore because you must exhibit the subset in question and then count it. *Natural,* or created structure (graphical, combinatorial, algebraic, or some combination of the three), enables one to count the set without iterating the set. For example, how many permutations are there of three symbols? Easy, list them 1,2,3; 1,3,2; 2,1,3; 2,3,1; 3,1,2: 3,2,1 and count them. How many permutations are there of 52 symbols? Well, the functional structure of permutations enables one to create the structured count 52! without listing anything. A more complicated example is to measure how long it takes to shuffle a deck of cards. Indeed, you can't define precisely what you mean by the question until you impose group structure on the set of permutations that model the shuffling process. And that is a reason, among others, that I start discrete mathematics with a two-day class discussion about what it means to shuffle a deck of cards.
>
> Once you start counting in structure you almost always discover that your answer depends on some hidden or confusing or implicit equivalence relation. How many permutations are

there of three symbols? Six you say? Not hardly. Any "sophisticated" mathematician will tell you there are only three: id, (a,b), (a,b,c); i.e., the three are: the identity, a transposition and a three-cycle, because there isn't a dime's worth of structural difference between the transpositions (1,2), (1,3), and (2,3) nor is there a dime's worth of difference between the 3-cycles (1,2,3) and (1,3,2). If you care only about parity, there are only two permutations (of any given degree) —an even one (id, or (1,2,3) or (1,3,2)) and an odd one ((1,2) or (1,3) or (2,3)). And on and on. What's really going on in any equivalence relation is that the equivalence classes are just the orbits of some group of permutations acting on the universe in question. (Sometimes this is of practical use and sometimes it isn't, but it is always in the background.) For example, (1,2) and (1,3) are naturally equivalent because they are conjugate. That is, the symmetric group on three symbols (S_3) acts on itself by conjugation ($xg = g \times g^{-1}$) and under this action, really a group of permutations of S_3, (1,2) and (1,3) are in the same orbit because $(1,2) = (2,3)(1,3)(2,3)^{-1}$. A way to look at this: to invoke the transposition (1,2)—don't! Rather go someplace (= (2,3)), do something else (= (1,3)), come back (= $(2,3)^{-1}$). Indeed the mantra all students should be reminded of incessantly is "go someplace, do something else, come back"—the essence of transformation and in some sense, for me at least, the most important idea of theoretical and applied mathematics. Never solve a hard problem, find a point of view which makes the problems solution conceptually obvious: don't multiply, add; don't use that messy matrix, diagonalize it; don't convolve, use an inner product and on and on.

And of course what's really behind my "go someplace, do something else, come back" mantra (modulo some algebraic technicalities) is the notion of commutativity (and we are back to algebraic structure); i.e., is AB = BA, where this equation is to be taken with a grain of salt (here is where I'm moding out the algebraic technicalities associated with unary functions and binary functions for the sake of pedagogical impact). Wouldn't it be wonderful if we lived in the commutative world all beginning undergraduates do? They are trained to mindlessly invoke "computational tricks" like $a + b = b + a$ and $ab = ba$ until they come to believe that

$$\text{flarn clarp} = \text{clarp flarn}$$

is a truism to invoke, rather than a question to ask: Is the flarn of a clarp the clarp of the flarns?

Is a root of a sum the sum of the roots? Is the power of a sum the sum of the powers (in particular, is the reciprocal of a sum the sum of the reciprocals?)? Is the integral of a sum the sum of the integrals? Is the derivative of a product the product of the derivatives? Is the log of a product the product of the logs? When the answer is yes, we are in high mathematical clover. When the answer is no, sometimes we are stuck—but sometimes we are actually in higher mathematical clover because we can find a tworble so that

$$\text{flarn clarp} = \text{tworble flarn};$$

$$\text{i.e.,} \qquad \text{clarp} = \text{flarn}^{-1}\ \text{tworble flarn} \qquad \text{or} \qquad \text{flarn clarp flarn}^{-1} = \text{tworble}$$

and the conjugation is less expensive than the original computation. So, the mantra "go someplace, do something else, come back." The students have a history with this idea (although they don't realize it without some serious provocation):

$$\text{log product} \neq \text{product log}$$

but

$$\text{log product} = \text{sum log}$$

so

$$\text{product} = \text{log}^{-1}\ \text{sum log}.$$

(A notational comment: to make the mantra work you must read the right-hand side from right-to-left because of the analyst's penchant for writing functions to the left of their argument.)

Current two-year curricula in mathematics emphasize calculus, differential equations and, perhaps, statistics. As a result, students often complete two years of intensive, high-quality mathematics with, at best, minimal knowledge of mathematics that has been developed in the past hundred years. Where are the powerful mathematical ideas that have been developed in the twentieth century? In number theory? In algebra? In quantum mechanics? Which of these ideas will increase the capacity of students to learn and create? How are these ideas related to calculus, differential equations, and statistics? How can powerful new ideas in mathematics be merged with existing ideas in calculus, differential equations, and statistics to create a two-year curriculum that will increase the fluency of students in mathematics and increase their capacity to learn and create? Once a set of powerful ideas has been identified, next we need to describe a set of activities, which students sufficiently fluent with these ideas, will be able to master. Now that we have identified the outcomes for a two-year curriculum in mathematics, let's consider ordering topics.

Currently curricula in calculus, differential equations, and statistics emphasize a topical order that develops ideas in an order so that students do not have to use concepts that they have not explored thoroughly (although many will disagree with the adverb). So, concepts of limit and continuity must be introduced and explored before the derivative can be examined. Ordering mathematical topics via topical precedence is logical, but students may struggle cognitively and motivationally with where to attach concepts such as limits and continuity because they cannot see the big picture. Efforts in reform calculus have offered substantial changes in the order, but the emphasis remains on thorough understanding of prerequisite topics before introducing the next topic. However, after graduation, scientists, engineers and mathematicians often learn only fragments of an area in order to work on their current problem. In order to increase capacity to learn and create, it may be more desirable to order ideas around questions that could be attacked. The processes in which students participate can be as important or more important than the ideas that are presented to students. Let's consider curricula in which order is generated from the questions that students will be asked to engage.

People learn what they have opportunity to learn. If students work in a classroom environment where the teacher presents information about powerful ideas, offers examples, and provides straightforward homework and examination activities based on the examples, then students will learn to do straightforward activities. In general, they don't learn mathematics, science, and engineering: posing problems, thinking about different ways to approach a problem, discovering connections, because they aren't doing mathematics, science, and engineering. Offering challenging homework and examination activities is an inadequate substitute, because students may have backgrounds in which they have not tackled a sufficient number of these types of challenging activities. Therefore, they are unsure about how to proceed, can become easily frustrated, and fail to take advantage of the opportunity to learn. Once we have our powerful ideas, let's consider carefully questions that can be posed in class to engage students in grappling with these powerful ideas. However, these questions must be developed carefully. "The student must find the question natural and interesting. If you find the question natural and interesting, they might and if you don't, they won't."[1] Ultimately, what students do is what they will learn. If they participate in the process of tackling interesting questions and applying powerful mathematical ideas, then they will increase their capacity to learn and create.

What are natural and interesting questions? Given the price of stamps since 1900, what is your prediction for the price of stamps in 2015 and how do you support your conclusion? Given two interacting species, how do populations of the species vary over time? What is a model for the data? What patterns are present in a set of data? How fast is the car traveling? How do you build a guidance system? What is a magnetic field? How much energy does it take to add two one-bit numbers? What is the age of the universe? What does a filter do to sound? How does a computer work? I'm sure you can devise better questions than my short list.

[1] Gary Sherman, Department of Mathematics, Rose-Hulman Institute of Technology, private communication.

What do we include? Within a limited time, we include a set of topics that maximizes the increase in the capacity of graduates to learn and create. What are the topics that provide the largest increase in the capacity to learn and create? Our experience suggests that the answer is those topics that promote the ability of students to quantitatively describe, understand, and predict behavior in the world around them. It is also suggested that what is required is a synthesis of ideas that include twentieth-century mathematics. Don't be limited to the currently accepted silos of calculus, physics, differential equations, algebra, number theory, chemistry, statistics, biology, linear algebra, etc. Are these silos important? Yes, but science and engineering students don't have time to take traditional courses in all these areas. Do we just sigh and say, "But that is how the material has to be arranged, because that is how the material is currently arranged?" Science and mathematics students will have to make do with small portions of what could be offered. Or do we think hard about the wealth of mathematical ideas and their relationships to behavior in the world around us and synthesize a set of powerful ideas that fit in the allocated time? How do you present these topics? We synthesize sequences of natural and interesting questions to engage students in doing mathematics. To the questions posed about the goals and content of mathematics, three answers have been offered: 1) focus on increasing the capacity to learn and create, 2) synthesize ideas to build new, powerful ideas that increase the capacity to learn and create, and 3) develop sequences of natural and interesting questions to engage students.

Goals and Content of a Two-Year Mathematics Curriculum

Frank Giordano
COMAP

Abstract. This paper argues that the question of content for our math programs is not nearly as important as the question of how we track the individual development of our students through the program. Student growth has to be the touchstone goal for our core programs. First we give some structure to the idea of students growing throughout our programs, and describe one way of looking at the dimensions that this entails. We present a reasoned choice of program content that can serve as a vehicle for student growth. Finally we examine the role of the instructor in planning for and building in opportunities and expectations for student growth throughout the program.

Overview

During the first two years: *What shall we teach and how shall we teach it?* Two years is a significant amount of time in the development of a student. At the end of the sophomore year, will students have progressed sufficiently to read and understand scientific exposition independently? Will they be able to research and learn mathematics on their own?

The first two years of college represent a crucial stage in the growth of a student. During this period the student is expected to transition from the high school classroom to the environment of upper divisional courses in applied science and engineering, where typically they are expected to learn and apply relatively sophisticated mathematical exposition as an implied part of learning science and engineering. Further, they are expected to assimilate the information in an increasingly independent fashion. For most science, mathematics, and engineering students, the only department teaching in each of the first four semesters is mathematics. Obviously, the content of these four courses is important if the courses are to provide the necessary mathematical foundation. But perhaps more important is the role these courses play in the development of a life-long learner. How does an educator provide for the progressive development of a student during the first two years?

Student Growth Dimensions

Among the areas needing careful attention during the first two years are:

Learning how to learn: Mathematics is the language of science. My experience is that few freshmen come prepared to read mathematics independently. Yet, beginning sophomore year or earlier, other disciplines will rightfully assume a proficiency in reading and writing mathematics as an outcome of the student's previous mathematical preparation.

Communication: Key to mastering the language of mathematics is expressing oneself comfortably in the language, not merely regurgitating definitions, theorems, and formulae that have often been memorized with little real understanding.

Mathematical Sophistication: Critical to a student's progressive development in mathematics is a curriculum organized so that it progresses in degree of sophistication of the underlying ideas. This is not simply a question of prerequisites – some ideas are harder for students to master. If students are to learn in an increasingly independent manner, then attention must be paid to the ordering of the topics. For example, the simpler ideas of matrix algebra may be introduced early in the curriculum, while the more sophisticated ideas of linear algebra are delayed. Similarly, if students are not comfortable with the notion of a finite difference, it is not likely they will understand the concept of a limit of a difference ratio.

Modeling: Students gain greater confidence and appreciation of applied mathematics if they can use the mathematics they are learning to approximate the world about them. Gaining an understanding of how the modeling process facilitates the use of mathematics to represent rather complex behavior is excellent preparation to read and understand scientific exposition. Further, modeling can be used to motivate the learning of new mathematics as the curriculum unfolds, connecting what otherwise may appear to the student to be disparate mathematics. For example as the curriculum unfolds, simple linear, deterministic, discrete models can be refined to consider nonlinear, continuous, and probabilistic considerations as appropriate.

Technology: The first two years provide an opportunity to prepare students to solve interesting applications using technology. If designed properly, this experience can shed greater understanding on the underlying mathematical concepts while enabling students to solve problems not tractable without technology. Mathematics can be more exciting if technology removes some of the tedium often associated with courses such as calculus. Finally, used creatively, the visualization of key mathematical concepts can be enhanced remarkably with technology.

Connectivity: The mathematics presented during the first two years should reveal connections to the larger curriculum. Most students want to know how the disciplines are connected, especially in courses they are concurrently taking.

The History of Mathematics: Many students find that learning the *who, why, when*, and *how* as well as the *what* in the creation of mathematics is interesting, and often motivational. For some students, it makes mathematics come alive. Students should understand that mathematics is an ongoing human endeavor. Reading and research projects in the history of mathematics provide further opportunities for students to practice communication skills.

Content Objectives

The needs of all disciplines in the overall curriculum are important considerations in the development of the two-year mathematics curriculum. But providing the opportunity to present the selected concepts in a coherent and mathematically sound way is crucial to the ultimate design of an experience that permits progressive student growth in each of the growth dimensions enumerated above. Given the opportunity to examine the tradeoffs inherent in designing the two-year experience, I believe each institution would design distinct curricula. For the purpose of discussing how to provide growth opportunities, the following content objectives are offered. The choice is based on a desire to present a *breadth* of applicable mathematics: discrete and continuous; deterministic and stochastic; and linear and nonlinear. The objectives selected for each of the following topics are based upon a hypothetical 15-hour curriculum, and my understanding of numerous "give and take" discussions among mathematicians, scientists, and engineers over the years. Such discussions are necessary for a core mathematics experience that is both useful as preparation for other disciplines and mathematically sound. Participation by other disciplines in the design of the core mathematics experience increases the probability that key mathematical concepts will be reinforced throughout the curriculum while fostering a sense of ownership and cooperation across the faculty.

Discrete Mathematics: Difference equations through systems of difference equations. An ideal subject for integrating modeling, technology, and contemporary mathematics building upon high school mathematics. Furthermore, an understanding of finite differences and finite sums is an excellent preparation for understanding calculus.

Matrix Algebra: Ideally, matrix algebra through the eigenvalue-eigenvector problem can be integrated early to be used immediately to solve systems of difference (or differential) equations. The higher concepts of linear algebra can be revisited later in the curriculum.

Calculus: Through multivariable calculus for all students. The end goals for coverage of vector integral calculus may vary among the particular disciplines being serviced by the core curriculum. Depending on when majors are declared, some tailoring of the multivariable curriculum may be possible.

Differential Equations: First and second order constant coefficients with the types of forcing functions normally seen at the undergraduate level. If matrix algebra and systems of difference equations are presented, a study of systems of differential equations extends the student's power to model real behavior while reinforcing important fundamental mathematical concepts.

Probability and Statistics: An introductory experience in probability and statistics is essential to provide a broad overview of the mathematics we use to model our world. Minimally, the concepts of the enumeration of sample spaces, conditional probability, independent events, expected values, and decision making under uncertainty should be included. A brief introduction to descriptive statistics and curve fitting should also be part of the introductory curriculum. Finally, a brief introduction to Monte-Carlo simulation enhances immensely the student's ability to incorporate the element of chance in their analysis, while aiding the understanding of probability.

Model Curricula

Examples of how several schools have selected combinations of each of the above content areas while integrating student growth objectives appropriate to their institutions occur at West Point, Carroll College, and Harvey Mudd College. Each two-year curriculum differs in the final objectives for each of the content areas, and the amount of "weaving" or "revisiting" of the content areas.

Providing Growth Opportunities

Student growth in each of the growth dimensions can be planned for whatever mathematical content is finally selected for the first two years. For each dimension, it is useful to select reasonable "end outcomes" and then select achievable and measurable intermediate objectives that occur as the curriculum unfolds. Viewed in this manner, the objectives for each dimension are planned "down" the curriculum. The curriculum can be viewed as a matrix, or fabric, with the growth objectives woven with the content dimensions. For example, in the modeling dimension, one may select final objectives that include building models of behavior in discrete and continuous environments. Eventually, some of the models may need to be refined to consider chance, or perhaps a more accurate model that employs nonlinear mathematics requiring numerical approximation. If the first course treats difference equations, the concept of proportionality can be reviewed as elementary constant coefficient difference equations are built. At that point problems such as pollution with dumping occurring at discrete intervals can be modeled. The models can be refined to include connected multiple pollution sites when systems of difference equations are studied. As the curriculum progresses to calculus, the question of continuous dumping can be treated using differential equations and systems of differential equations. Students see that the mathematics they are learning extends their ability to model the world about us. Such an approach provides additional threads to connect the curriculum internally (the mathematics curriculum) and externally (the other disciplines). The external connection can be further enhanced with input, and perhaps participation, in the design, presentation, and critique of the modeling projects. Finally, the projects provide opportunities in communication, technology and the history of mathematics. A similar approach would be followed for each of the other threads.

The Role of the Instructor

As the above curriculum unfolds, the role of the instructor changes. Perhaps as a transition from the high school environment, the instructor initially is like a "coach", providing the initial mathematical "training" while urging students to greater achievement. Quickly, however, the need for counseling becomes greater as the instructor must continually assess every student's status in each of the growth dimensions and design opportunities for students to take their next step. Later in the curriculum the instructor becomes more of a "travel guide" than coach, pointing out the interesting places to visit as the students begin to learn more independently. If we ask the students to learn in an increasingly independent manner, than we must provide the motivation to do so. What motivates a particular student and how do you "hook" them—is it history, modeling, mathematical rigor, using technology, or connections to other disciplines? While perhaps not as glamorous as giving an eloquent lecture or being the source of knowledge in the classroom, guiding student growth may be more beneficial to most students in the long run.

Instructional Techniques Perspective

This section edited by Kathleen Snook.

The calculus reform movement called for a shift of focus from teacher-centered to learner-centered in the pedagogical arena and less coverage and greater depth in the content arena. The pedagogical shift involved engaging students in multiple learning activities such as group activities, group projects, discovery work, calculator or computer laboratory sessions, writing assignments, and student presentations. Recently there has been a greater emphasis on inquiry and modeling. The benefit of each of these learning activities is established in the research literature and each has its own support group of instructors. The difficulties in obtaining widespread adoption lie in the fact that student-learning activities take class time, require more work for the instructor (compared to the lecture format), and require a change in assessment methods. The time factor poses the biggest challenge. The undergraduate teaching profession has been reluctant to reduce content in order to make time for student learning activities and have difficulty accepting that learning is a very inefficient process.

Kathleen Snook writes "Constructivism has been the major guiding philosophical basis for pedagogical change within mathematics classrooms in the last decade." With constructivism as a background Snook discusses nine instructional methods: Questioning/Discussion, Problem Solving, Use of Technology, Exploration and Discovery, Multiple Representation, Writing, Multiple Assessment Instruments, Section Size, and Use of Group Work.

Elizabeth Teles describes the three top educational achievements recognized by the National Science Foundation's Division of Undergraduate Education during the twentieth century:

1. Paradigm shift from teacher-centered to learner-centered education.
2. Recognition of the need to balance learning of facts and learning of process.
3. Exploitation of technology for teaching and learning.

Teles notes the parallel pedagogical recommendations of the American Mathematical Association of Two-Year Colleges, the calculus reform movement, and science, and then addresses the implementation question.

Shirley Pomeranz draws together a variety of research and writings that address the major issues and future directions of teaching undergraduate mathematics. Pomeranz echoes the belief that although techniques that result in student-centered active classrooms are effective for learning, instructors have difficulty finding the time required for proper planning and implementation of these techniques.

Chris Arney writes about using mathematics, its process and its pedagogy, to enhance and develop creativity, one of the foremost needs of society as we look to the future. Stating that "Creativity is fundamental to the development of productive graduates," Arney explores elements in a core mathematics program that lend themselves to instructors building the ingredients of creativity in their students.

A Continuum of Choice: Instructional Techniques in Undergraduate Mathematics

Kathleen G. Snook
United States Military Academy

Abstract. Constructivism has offered a philosophical basis for change in the teaching of mathematics. With the belief that all students construct their own knowledge and do so in very different and individual ways, it becomes imperative to use a variety of teaching strategies. This paper considers teaching strategies that can be implemented along a continuum of choice to move toward a student-centered *constuctivist* classroom.

Introduction

In 1995, when doing an initial assessment of the first decade of calculus reform, Leitzel reported that although some content in calculus courses had changed, the more apparent and widespread change was pedagogical [1]. Instructors were making changes in undergraduate mathematics' classrooms in an effort to increase the emphasis on conceptual understanding and decrease emphasis on symbol manipulation. Many schools were making use of graphing calculators and incorporating modeling and applications into their courses. Some schools began to replace the traditional undergraduate large-section mathematics lecture class with smaller sections in which students experienced one or more of a variety of instructional practices. In recent years, pedagogical changes have continued in undergraduate classrooms.

Constructivism has been the major guiding philosophical basis for pedagogical change within mathematics classrooms in the last decade. One can trace the roots of constructivist ideas to general research in cognitive psychology done in the latter half of this century, and more specifically to the work of Piaget. Piaget believed that action and knowledge are inextricably linked [2]. Constructivists generally agree that all knowledge is constructed by the learner, and that cognitive structures are under continual development. Constructivists believe that it is purposive activity that induces the transformation of existing structures, and that the environment presses learners to adapt [3]. Once an instructor acknowledges a constructivist perspective as a cognitive position, methodological constructivism follows. "Once a constructivist perspective is adopted, the day-to-day life in the classroom is profoundly and significantly altered for both teacher and students [4, p. 314]." One does not, however, need to understand the deeper cognitive aspects of constructivism to teach in what would be considered a constructivist manner. Many teachers are natural constructivists, guided by what they feel is most beneficial for their students: a more interactive and engaging classroom.

Through the constructivist lens we see an obvious distinction between lecture or traditional teacher-centered methods and active learning or student-centered methods. One can picture a continuum of classroom environments in which a variety of methods reside. On the one extreme there are classrooms in which only the teacher speaks or lectures. On the other extreme, a teacher may not even be present as the students independently and individually work through the material. (I propose neither of these extremes used exclusively serve the students well.) As one moves from the lecture to the independent study, varying degrees of action by, and interaction between, teachers and students occur. Methods which may influence movement along the continuum include the use of questioning or discussion, (applied) problem solving, technology, explorations and discovery, multiple representations of mathematics, writing, various types of assessment instruments, smaller section sizes, and collaborative or cooperative groups. Gradations of student-teacher action and interaction occur within these methods spreading various adaptations of each along the continuum as well. In using a continuum image, instructors can select general strategies or methods and implement them within their comfort level. Although there may be additional methods, I will consider the nine listed above in addressing some of the issues and future direction of instructional methods in undergraduate mathematics.

Instructional Methods: Strengths and Drawbacks

Questioning/Discussion. The first step toward a more active and interactive classroom is to let students enter the mathematics discourse. We can accomplish this, for example, by the instructor questioning students (Socratic type interaction), by the instructor turning students' questions back to other students, or by the instructor allowing the students' questions or discussion to control the flow of the class. Although varying in their degree of being student- or teacher-centered, these examples illustrate that in place of the instructor doing all of the speaking, he or she facilitates the students in actually presenting the material. The advantage of this method is that students are forced to articulate ideas in their own words. Other students may understand these explanations better. The student who is speaking is not only required to think about, or perhaps even mentally image, the idea, but also must verbalize it. Student talk is very revealing of students' understanding [5]. Instructors can learn a great deal from what a student says. In this type of classroom students may arrive more prepared if they know the instructor expects some degree of participation.

Some instructors voice concern about opening one's classroom to discussion and questioning as it may take control away from the instructor. Allowing student participation in classroom discourse takes more time and class discussions may head down a path not planned by the instructor. Many teachers consider this a strength as students take more responsibility for their learning. These discussions may lead to less material being covered than in a traditional lecture. In this environment, instructors must pay very close attention to what is being said and how it is said. Poor articulation by one student can lead to the development of misconceptions by other students.

Problem Solving. Students are interested in the worth of the mathematics they are studying. The mathematics studied by undergraduates in their first two years provides a bridge between high school algebra and engineering and science. Problem-solving activities, and especially real-world applications, provide students a picture of how the mathematics they are learning is used. Toward one end of our continuum is supplemental problem solving. Textbooks usually provide some supplemental problem solving activities in the list of problems at the end of each section. These begin to help students see the uses of the current topic under study. A step further using this method is to integrate applications into the course. Application problems can serve as the vehicles for learning concepts. Real world scenarios with realistic data often motivate students. Real-world problems offer some degree of uncertainty from which instructors can discuss the concepts of assumptions and accuracy, as well as the appropriateness of various mathematical tools. As one moves toward more student autonomy, instructors can use interdisciplinary projects to provide a realistic and fully integrated problem-solving approach. These type projects require students to model situations from another discipline mathematically, choose from an array of solution techniques and translate their solution to address the original problem. These projects benefit both students and teachers in learning more about the use of mathematics by other disciplines.

The largest difficulty in incorporating problem solving is time. Students must be given time to wrestle with uncertain situations. Although having students involved in the modeling process is very beneficial, when integrating applied and interdisciplinary problem solving into a course, time must be allocated for this endeavor. Instructors should give compensatory time to students for major project submissions. In making a decision to truly integrate problem solving, one must make the difficult decision of what to delete from the course syllabus. Additionally, developing real-world interdisciplinary projects is no small undertaking for faculty members. Coordination between faculty in partner disciplines is professionally worthwhile, but also time consuming.

Technology. Many calculus reform projects focused on technology use. These projects largely looked at integration of calculators and computer algebra systems (CAS) into calculus courses. Instructors looked at technology to provide the time needed for other activities through quicker calculations and graphing. Instructors have incorporated technology in varying degrees. On the one hand, instructors can simply allow calculator use in the classroom, while on the other they can demand every student own and use a particular calculator or CAS. In some cases instructors are teaching/facilitating courses exclusively through or with a computer (see Exploration and Discovery method). Technology has allowed a refocusing on conceptual understanding versus procedural proficiency. Students can develop a model or set up a problem and then use their hand-held calculator or CAS to solve the model. Hand-held technology offers quick and easy generation of numerical data and graphs in the classroom. Students can perform symbolic manipulation using technology rather than a pencil. Calculators now provide many symbolic analytic solutions. Instead of spending time generating data, plotting points, or performing algebraic manipulations, students can spend time analyzing and understanding the model phenomena. Both the current available calculators and CAS provide essential support to students' applied problem-solving efforts discussed above. Technology offers students the ability to explore and discover.

Concerns over fundamental skills linger. What should students be able to do without a calculator? What is the importance of hand calculations/skills? Which concepts/skills should be learned first by hand, but once understood shifted to a calculator procedure? When do we introduce CAS? With current calculators, when do students really need a separate CAS on a computer? How do we make available technology a tool and not a crutch? In only a few short years our students will be the middle school children of today who have grown up with technology. Their view and use of technology will be very different than that of even today's undergraduate. The mathematics community must think deeply about these questions as we prepare to educate these students in the twenty-first century.

Explorations and Discovery/Facilitated Laboratory. In some classrooms technology is added on to the course curriculum. In laboratory classrooms technology is integral to the course. Examples of laboratory type courses include Calculus Using *Mathematica*; Calculus, Concepts, Computers and Cooperative Learning (C4L); Calculus and *Mathematica*; Calculus in Context; and Project CALC: Calculus as a Laboratory Course [6]. Those involved in the development of these courses believe that technology is not a solution to pedagogical problems, but offers alternative choices to address those problems. A comparative study between a traditional course and a Calculus and *Mathematica* (C&M) course at the University of Illinois showed the C&M students obtained a higher level of conceptual understanding while maintaining computational proficiency, and also showed a more positive disposition toward mathematics and computers [7]. In laboratory courses students discover mathematics through exploration, conjecture and verification using technology. They also program computers to perform procedures. In doing so, students develop understanding of the processes involved. Researchers in the C&M study concluded students benefited by better visualization of ideas which promoted sound conceptual understanding, discovery learning by exploration which induced reflection and resulted in developing relationships between concepts, and cooperative learning which established rapport and teamwork among the students [7].

One of the concerns of laboratory courses is that students will blindly use the technology without understanding the concepts (see Technology). This is preventable if the laboratory course is carefully designed and facilitated. Assessment instruments must be designed that measure both conceptual and procedural understanding, not just technological proficiency. Laboratory courses also are time consuming. Discovery and exploration require time for conjecturing, testing, reformulating and reflecting. A final concern for some schools is that computer laboratory facilities necessary to offer a laboratory course are not available. The alternative, to have students purchase laptop computers, may also not be feasible.

Multiple Representations. The "rule of three" was introduced during calculus reform efforts based on the belief that emphasis should be placed on numerical, graphical, and analytical representations of mathematics. Since its initial introduction some educators have moved to a "rule of four" to include the representation of mathematics in words (either orally or in writing). Focusing on various representations assists students in understanding and making connections between concepts. Additionally, these representations provide students with weak algebra/symbol manipulation skills the opportunity to grasp the concepts while they hone their procedural skills. Student participation in the classroom and writing in mathematics support the concept of representing mathematics with words. Technology has opened the floodgates for various representations of mathematics. Students can quickly obtain a graphical or numerical representation of a particular function. They can begin to analyze its behavior immediately. Students can observe and describe long-term behavior and examine limiting processes. CAS has provided analytic support for the student and made it easy to examine simultaneously the analytical, numerical and graphical representations of problems and solutions.

There are very few drawbacks to incorporating multiple representations of mathematics in the classroom. One, however, is students' acceptance of one representation as "proof." As core or service course mathematics classrooms have shifted away from formal proofs, some students may believe all representation are equal and will use one example, or in this case one representation, to generalize or draw a conclusion. For example, a student's conclusion from analysis of a particular graphical solution may be incorrect due to the domain in which the graph was viewed. Additionally, many times in undergraduate mathematics we do not have graphical or numerical solutions, but truly have graphical or numerical *approximations* to solutions.

Writing. More than just a fourth representation of mathematics, student writing truly reveals students' understanding. Having students articulate their interpretation and analysis of mathematical concepts and problems is extremely revealing. In turn this provides feedback to the teacher for instructional responses. Writing prepares students to clearly communicate mathematics to clients or outside reviewers of their work. It forces logical and organized thought. One can integrate writing into a course in varying degrees. Exams or quizzes can include short answer or explanation type problems. Given a scenario and a graph, for example, students can describe the link between the behavior of the graph and the scenario. After using an approximation technique, students can discuss whether or not their solution is reasonable. Instructors can assign essays that require students to explain a course concept to perhaps a friend who is not taking the course. The use of journals has become more common in mathematics courses. Student journals may include concept analyses, reflective summaries, annotated problem solving or essential term definitions.

Developing good writing requirements is difficult, and assessing written work is equally as difficult for most beginners in this area. We, as well as our students, will need time to develop as we incorporate writing into mathematics. Time again becomes an issue to consider. Just as various types of analytic problems require varying time efforts from students, time required for writing varies with the requirement. All writing requires some time for reflecting and organizing one's thoughts before composing the prose. Assessing writing requirements also involves instructor time; we need to carefully listen to (read) what our students are saying.

Assessment Instruments. As we change how and what we teach, we must change how and what we assess. With the advent of technology, a proficient calculator user can pass many traditional exams without having an understanding of the material. The reasons given above to support various instructional methods apply to support varying assessments. Assessments can include a combination of modeling, problem solving, writing, producing or analyzing multiple representations, technology, analytic calculations, or symbolic manipulation. They can be in the form of quizzes, exams, essays, projects, problem sets, or journals. Assessments should reflect the methods and approaches the instructor has used in the classroom. Results from assessment plans that use a variety of problem types and problem presentations offer a more comprehensive view of students' understanding [5].

Developing effective assessment instruments that include various types and presentation of problems is difficult, but certainly possible. Instructors must be careful about how much they ask students to do especially on time sensitive exams or quizzes. When students are asked to explain a concept or model a situation, these requirements take longer than simply, for example, calculating a derivative using the chain rule.

Small Section Sizes. As one reads about the benefits from the possible instructional techniques above, it becomes evident many would be difficult to implement in a classroom with a large number of students. One indirect result of the methods above is that as more interactions occur between teacher and student, the teacher becomes more familiar with the students. Within smaller sections, the teacher learns more about the levels of students' conceptions, as well as their misconceptions. Assessment (not necessarily evaluation) occurs daily within the classroom activities. Additionally, in small sections teachers tend to do more of their own grading which offers a more revealing view of the students. With a manageable number of students, the teacher can then mentor these students of mathematics in an informed manner and better facilitate their learning.

Large sections of mathematics courses are certainly economically beneficial. Many students can pass through the course and the cost is only a few instructors and perhaps a few more teaching assistants. Large sections tend to enlist the assistance of graders and teaching assistants which reduces the administrative and tutoring burden on faculty. At some locations resources other than faculty and budget prescribe section sizes. These may include limited facilities and scheduling issues.

Cooperative and Collaborative Groups. Whether restricted by large section sizes or working in smaller section classrooms, cooperative and collaborative groups can assist in more interactions between teachers and students. While both types of groups indicate students working together, some researchers and teachers make the distinction between cooperative group members having different roles, and these various players cooperate to solve problems, while collaborative group members are truly working together as one entity toward a problem solution. Group work facilitates students discussing the mathematics at hand in their own words. They must verbalize and explain their reasoning to peers. This, in turn, organizes their conceptions for improved understanding. Additionally, group and teamwork experiences are becoming essential for future study and employment. Group work has been successful in many settings, to include large classrooms [8] and supplemental instructional (SI) programs (see underrepresented groups) [9].

One must be cautioned that within group discussions and problem solving, misconceptions may be formulated. Feedback loops must be established in an effort to prevent this occurrence. When assigning group requirements, some members may not be as engaged as others. Although mirroring possible situations in future courses or employment that these students will have to face, some undergraduates are ill equipped to deal with the non-participatory group member. Caution should be taken when requiring a group submission. Even when all members are engaged, preparation of the final submission usually falls on one or two members of a group. Frustration with group work may disengage some students from the experience.

What methods best increase success of underrepresented groups?

The largest factor in improving success of underrepresented groups may be active engagement. By using some of the techniques described above, students become engaged in learning mathematics. "Windows into mathematical thinking" are opened to previously excluded groups through alternatives to algebraic manipulation, exploration of ideas from multiple perspectives, active classroom participation and a cooperative rather than competitive environment [10]. Programs that "nurture confidence and build stronger study skills as students learn" are needed for these groups [11]. In these types of programs students have mathematical experiences that prepare them to use and understand the mathematics they have studied. A key ingredient to the effectiveness of these programs is equity of access.

At sites where reform has been adopted (or adapted), students are at least as well prepared mathematically as they have been previously and have had a more comprehensive experience [1, 12]. At some sites, programs outside of the regular classroom are offered to support all students. Support activities include mentoring and tutoring. These programs have a significant impact on underrepresented groups. One such program is Supplemental Instruction, a model of learning assistance offered outside of the targeted class [9]. Perhaps these engaging environments and supplemental programs will motivate a number of students to continue their study of mathematics.

How to choose and integrate various instructional methods?

In choosing instructional methods one must consider both personal comfort and the environment in which they are working. The methods described above can be used intermittently as a supplemental part of a course, or daily as an integral part of a course. If we wait to be completely comfortable with a method before we try it, we might never attempt innovation. On the other hand, we should work within our comfort levels. Many times it's easier to start with small changes and then attempt larger endeavors.

The environment can have a significant impact on what one is able to do in the classroom. For example, some methods are difficult to do, although not impossible, in large sections. Time constraints on students or faculty may limit problem-solving activities. Classes at schools where most students are commuters may find it difficult to assign group projects. Technology may not be available in the form of computers, so calculators may have to suffice. Appropriate instructional methods should be chosen to fit the environment. Just as in life, any variety is bound to spice up your mathematics class.

Bibliography

1. J.R.C. Leitzel, "ACRE: Assessing calculus reform efforts," *UME Trends, Vol. 6*, No. 6, (January 1995) 12–13.
2. J. Piaget, 1971. *Psychology and Epistemology*, (A. Rosin, Trans.) New York: Grossman Publishers.
3. N. Noddings, "Constructivism in mathematics education," *Journal for Research in Mathematics Education (JRME), Monograph Number 4, Constructivist Views on the Teaching and Learning of Mathematics*, (1990) 7–18.
4. M.L. Connell, "Technology in constructivist mathematics classrooms," *Journal of Computers in Mathematics and Science Teaching, Vol. 17*, No. 4, (1998) 311–338.
5. K.G. Snook, 1997. *An investigation of first year calculus students' understanding of the derivative*, Unpublished doctoral dissertation, Boston University.
6. K. Stroyan, D. Mathews, J. Uhl, L. Senechal & D. Smith, "Computers in calculus reform," D. Smith ed.), *UME Trends, Vol. 6*, No. 6, (January 1995) 14–15, 31.
7. K. Park & K.J. Travers, "A comparative study of a computer-based and a standard college first-year calculus course," *Research in Collegiate Mathematics Education II, CBMS Issues in Mathematics Education, Vol. 6*, (1996) 155–176.
8. R. Maher, "Small groups for general student audiences – 2," *Problems, Resources, and Issues in Mathematics Undergraduate Studies (PRIMUS), Vole VII*, No. 3, (September 1998) 265–275.
9. S.L. Burmeister, P.A. Kenney & D.L. Nice, "Analysis of effectiveness of supplemental instruction (SI) sessions for college algebra, calculus, and statistics," *Research in Collegiate Mathematics Education II, CBMS Issues in Mathematics Education, Vol. 6*, (1996) 145–154.

10. J. Ray, "What if the goal is not calculus? Implications of calculus reform for the two year college curriculum," *UME Trends, Vol. 6*, No. 6, (January 1995) 9–10, 31.

11. G.D. Foley & D.K. Ruch, "Calculus reform at comprehensive universities and Two-Year Colleges," *UME Trends, Vol. 6*, No. 6, (January 1995) 8–9.

12. J. Baxter, "Assessing reform calculus," *Focus on Calculus: A Newsletter for the Calculus Consortium Based at Harvard University, Issue No. 16*, (Winter 1999) 4.

13. J. Hiebert & P. Lefevre, "Conceptual and procedural knowledge in mathematics: An introductory analysis," *Conceptual and Procedural Knowledge: The Case of Mathematics* (J. Hiebert, Ed.), Hillside, NJ: Lawrence Erlbaum Associates (1986) 1–27.

14. L.A. Steen, "Reaching for science literacy," *Change*, (July/Aug 1991).

Shifts in Undergraduate Education: Learner-Centered, Process-Oriented, and Technology-Based

Elizabeth J. Teles*
National Science Foundation

Abstract. Program Directors in the Division of Undergraduate Education at the National Science Foundation identified the top three significant achievements in science, mathematics, engineering and technology education as a shift from teacher-centered to student-centered education, a focus on balancing the learning of facts and the learning of processes, and the exploitation of various technologies to improve teaching and learning. This paper explores how the mathematics community is currently involved in these activities and their related instructional strategies, and offers exciting and challenging prospects for undergraduate education in the twenty-first century.

Recently program directors at the National Science Foundation were asked to identify the most significant achievements in science, technology, engineering, and mathematics (STEM) research and education in the twentieth century and the rich areas anticipated for development in the twenty-first century. The top three educational achievements recognized by the Division of Undergraduate Education (DUE) have vital implications for instructional strategies that are employed in STEM education.

The first achievement is a paradigm shift from teacher-centered to learner-centered for undergraduate education. The difficult task of moving higher education to focus on what students actually learn fosters examination and re-formulation of the content and processes of undergraduate STEM education. The result has been a greater recognition by institutions of higher education that:

- instruction is enhanced by a research base;
- learning occurs in a large variety of curricular and extracurricular settings;
- sufficient motivation, introduction, and time allow for all students to learn complex concepts;
- information can be effectively presented in an integrated context which is interdisciplinary and tied to actual students' interests and real-world applications;
- assessment is a critical component of instruction; and
- faculty must be well-prepared for their roles as teachers.

The second achievement is the recognition that balance must be achieved between the learning of "facts" and the learning of processes. This is the true value of the integration of research and education.

---

* The opinions expressed in this paper are those of the author and are not intended to represent the policies or positions of the National Science Foundation.

Such learning must incorporate personal experience if it is to be effective. Students must have opportunities to see, hear, do, and teach.

The third achievement is the exploitation of various technologies (smart laboratory instruments, computers, calculators, modeling and visualization tools, the Internet) to allow students to:

- explore theories and concepts without getting bogged down in tedious calculations or manipulations; or
- learn outside the confines of a particular time and classroom or laboratory setting. This relates both to learning anytime and anywhere and to engaging students in the observation or simulation of processes normally too large, too small, too fast, too slow, or too dangerous for direct interaction.

How is the mathematics community involved in these activities and instructional strategies? How do these achievements intersect issues facing mathematics—particularly the way mathematics interacts with other disciplines? How will educational changes related to these achievements be supported by faculty, administrators, departments, institutions, and funding agencies?

The three achievements identified for undergraduate STEM education closely parallel the recommendations of *Crossroads in Mathematics* (the AMATYC Standards) and the goals and objectives of calculus and other mathematics reform texts. Proposals to DUE programs increasingly discuss not only content of the materials to be developed but also the instructional strategies to be used. Recently funded proposals include instructional strategies such as: (a) the incorporation of exercises that involve critical thinking and concept development, (b) cooperative activities, (c) the use of technology as a means of student construction and manipulation of concepts, (d) the use of reading and writing in addition to traditional mathematics skills, and (e) multiple representations of concepts.

The guidelines for pedagogy described in *Crossroads in Mathematics* [4] include active involvement of students, technology to aid in concept development, problem solving and multistep problems, mathematical reasoning, conceptual understanding, use of realistic problems, integrated curriculum developed in context, multiple approaches to problem solving, diverse and frequent assessment, open-ended problems, oral and written communications, and use of a variety of teaching strategies.

The article "Visions of Calculus" in *Calculus: The Dynamics of Change* [1] mentions several common instructional strategies in the reform calculus movement. These encompass multiple representations of concepts, the use of technology, student projects, writing about mathematics, and concept development. In *Assessing Calculus Reform Efforts* [2], Tucker and Leitzel note similar strategies and their impact on the teaching of calculus. The Statistical Abstract of *Undergraduate Programs in the Mathematical Sciences in the United States: Fall 1995 CBMS Survey* reports that almost 30% of calculus was being taught using "reform" texts; however, over 40% of teachers were using computer assignments and other "reform" instructional strategies. [3]

The article "Linking Teaching with Learning" in *Science Teaching Reconsidered* [5] makes a case for scientific research as a model for learning and teaching—specifically for *active* learning and *active* teaching as opposed to methods that college teachers have traditionally used, such as lectures, assigned readings, problem sets, and closely supervised laboratory work. The authors recommend methods such as engaging students in communities of learning, establishing a context for exploration, proposing explanations, and reading and writing for understanding. They give numerous research references that indicate that traditional methods are less effective than once thought in developing understanding.

Because of the widespread call for using alternative teaching strategies instead of depending almost totally on traditional lectures, there are—as there should be—frequent discussions of what students should know and be able to do. Most of the reports mentioned above recommend areas to de-emphasize and instructional strategies to use less frequently, as well as areas to emphasize and strategies to use. While there is some agreement that changes are needed, there is not at present universal agreement on what these should be. There has been some backlash against reform texts and methods of instruction. Movements such as "Back-to-Basics" even demand that high schools and colleges return to traditional methods of instruction. It is not uncommon to hear about "Math Wars" and lawsuits over the use of newer approaches.

What is needed to help college faculty and administrators decide what is best for students? One answer might be a high-level call within the mathematics community and related disciplines for the use of particular instructional strategies and knowledge. The Mathematical Association of America (MAA) has recently initiated a project entitled "Mathematics and Mathematical Sciences in 2010: What Should Graduates Know?" In this project, MAA hopes to develop in association with other disciplines a consensus on many issues. Workshops occurring in 1999 and 2000 at several locations are bringing together mathematicians and faculty from other disciplines. These meetings and workshops help to develop consensus and plans. However, in these efforts, care should be taken that participants are not just preaching to the choir.

Another answer may be more research and evidence showing that students learn more or can perform better with the newer methods and materials. One well-known researcher at a recent meeting, however, made the observation that it is primarily in the United States where there is a constant demand for more research, more proof, and more accountability. In most other industrialized countries, involved constituencies such as academia, government, and labor sit at a common table and look toward the future, and decide what educational practices will best meet national goals. Perhaps this is due to a more centralized system of education in other countries. It may also be because many citizens of the United States seem to look more to tradition and espouse the philosophy, "If it ain't broke, don't fix it," when it comes to educational reform and ignore the evidence that it may indeed be broken.

A third answer may be increased funding by college administrators, state and local funding agencies, business and industry, and government agencies such as NSF. Departments need increased funds for technology, professional development for faculty, student assistants, laboratory space, and other related items. While just devoting resources to a project will not assure success, the lack of resources can often doom it to failure.

A fourth potential answer is extensive support mechanisms and professional growth opportunities for faculty to assess and implement recently developed educational materials, emerging technologies, and teaching methods. [6] *Crossroads in Mathematics* [4] has a chapter on the implications of reform methods on faculty development, departmental considerations, advising and placement, laboratory and learning center facilities, technology, assessment of student outcomes, program evaluation, and articulation.

Looking towards the future, the following two issues were identified by DUE as exciting and challenging prospects for undergraduate education for the twenty-first century.

First, the increased recognition of the value of a research-base to the understanding of learning will lead to answers to six key questions:

- What are the critical factors, optimal environments, necessary boundary conditions, and resultant indicators for effective instruction within the various STEM disciplines?
- What are the social, cultural, and institutional factors that affect participation in STEM fields by individuals and demographic groups; and how are they mitigated or optimized?
- What are the effects of teaching and learning technologies on instruction, student learning, and student critical thinking?
- What are the indicators of success and attainment in STEM education and how are these correlated with input and output measures?
- What new pedagogic theories and techniques might be effectively employed within STEM disciplines?
- How will educational research inform education for adult and life-long STEM learning?

Second, the focus on educational complexity must be extended from pre-college to undergraduate education. Our nation's increasingly complex social systems, the increasing number of high school students who continue on to college, and the public desire to achieve accountability from a higher education system whose costs appear to spiral ever upward will result in the assignment of responsibilities that higher education has not previously had to assume. There will be greater demands for:

- the correlation of undergraduate education with employment outcomes, and

- social and support services as well as service-learning opportunities to provide greater attention to the affective component of learning, and coordination of these activities across academic departments, institutions, and levels (pre-college, two-year, baccalaureate, and graduate).

As we enter a new century, the time being dedicated to exploring where we have been and where we wish to go is a wise investment. The mathematics community should take the lead in articulating the major educational questions that need to be resolved and should do so in concert with other disciplines. Answers to these questions should be informed by past successes and a clear vision for the future.

References

1. Sharon Cutler Ross. "Visions of Calculus." In A. Wayne Roberts (ed.), *Calculus: The Dynamics of Change* (Mathematical Association of America, Washington, DC, 1996), pp. 8–15.
2. Alan C. Tucker and James R. C. Leitzel (eds.) 1995. *Assessing Calculus Reform Efforts*. Washington: Mathematical Association of America.
3. Don O. Loftgaarden, Donald C. Rung, and Ann E. Watkins (eds.) 1997. *Undergraduate Programs in the Mathematical Sciences in the United States: Fall 1995 CBMS Survey.* Washington: Mathematical Association of America.
4. Don Cohen (ed.) 1995. *Crossroads in Mathematics: Standards for Introductory College Mathematics Before Calculus,* Memphis: American Mathematical Association of Two-Year Colleges.
5. "Linking Teaching with Learning." In *Science Teaching Reconsidered* (National Academy Press, Washington, DC, 1997). pp. 21–27.
6. William E. Haver (ed.) 1998. *Calculus: Catalyzing a National Community for Reform.* Washington: Mathematical Association of America.

Major Issues and Future Directions of Undergraduate Mathematics: Instructional Techniques in Freshman and Sophomore Math Courses

Shirley Pomeranz
The University of Tulsa

Abstract. A variety of research and literature provide suggestions for instructional methods, as well as features of those methods. Even with an abundance of information available, instructors must still make choices of instructional strategies based on their own situations and environments. This paper discusses some of the possible choices by addressing benefits of the strategies and ideas about the targeted student population, the effects of the learning media, and the promotion of understanding.

Introduction

Major issues and future directions of instructional techniques in freshman and sophomore mathematics courses are briefly discussed in the course of considering the following eight questions.

1. What are the strengths/drawbacks of various instructional methods?
2. How should one choose and integrate various instructional methods?
3. Which methods best increase success of underrepresented groups?
4. How do the learning media affect reading, writing, and problem solving?
5. How should one build theoretical understanding?
6. How should one align the achieved curriculum and the intended curriculum?
7. What guiding principles arise from educational research?
8. Should calculus be a laboratory (discovery) course?

Strengths and drawbacks of various instructional methods

There exist a multitude of instructional methods available for use by instructors of mathematics. The Preface to *Calculus–Single and Multivariable*, Second Edition [1] recommends several general features that instructional methods should have:

- Focus on a small number of key concepts—emphasize depth rather than breadth of understanding.
- Encourage active learning.
- Employ multiple representations—geometric, numerical, analytical, and verbal.

- For example, represent functions graphically (picture), numerically (table of values), and algebraically (formula).

Instructional methods for freshman and sophomore mathematics courses include the following [2]:

- Traditional lectures.
- Use of computer algebra systems (CAS) and electronic notebooks in place of a traditional textbook.
- Pedagogical approaches based on a constructivist theoretical perspective of how mathematics is learned (described by a shift from a teacher-centered classroom, where lecture predominates, to a student-centered laboratory, where students make and test hypotheses and discover mathematical truths for themselves).
- Reform calculus, emphasizing real-world problems, hands-on activities, and discovery-based learning (involving conceptual problems as well as computation).
- In-class group-work activities, projects, and student oral presentations.
- Use of case studies.

The tradeoffs between these methods are explicitly given by their descriptions in the following sections.

Choosing and integrating various instructional methods

The term *calculus reform* has various interpretations. However, there seems to be a consensus that "calculus reform" includes active student learning in the classroom—students actively participate in their classroom activities, instead of assuming the traditional role as passive note-takers. Instructors present material from multiple perspectives in order to encourage understanding of concepts. Technology alleviates tedious computations and enables students to attack more real-world problems. Students can use technology to explore for themselves and to aid in visualization. Math diaries, group projects, and writing projects may also be components of a calculus reform classroom [3]. Issues about these methods involve to what degree each innovative aspect of calculus reform should be utilized in a specific course and by a specific instructor.

Technology, in the form of CAS, is now a component of most freshman and sophomore math courses. Current issues center on the degree of use of CAS. At one extreme, some courses are taught with computers in the classroom [4]. On the other hand, CAS may be used only as a supplementary tool to perform numerical computations and obtain graphics. One of the most touted benefits of CAS is its graphics capability and its use for visualization. This can be especially helpful in multivariable calculus [5]. In any event, the use of CAS opens a whole new range of pedagogical questions. What is the role of such technology? How are course objectives best met with CAS? What criteria are to be used in assessment of student performance? And there are many other related questions.

There are also various degrees of web-based courses in which the Internet becomes the source of an assortment of course material available to students [6], [7]. Use of the Internet can range from electronic instructor comments to help clarify and emphasize (or de-emphasize) specific material in the course text, to large-scale use of the Internet to deliver homework problems, assignments, and projects. Students may submit their work electronically. Distance learning is another example of an instructional technique involving the Internet (see Section 4).

Using case studies can also be an effective means to motivate engineering and physics students to appreciate the role of mathematics in engineering and physics disciplines [8], [9], [10]. Once students see the relevance of mathematics to their disciplines, there is interest and incentive to pursue mathematics.

There are many instructional methods that work in large lectures, small classes, computer labs, small groups, or traditional classrooms, and with various technologies (from graphing calculators to complete CAS). Various instructional methods should be chosen so as to fit the specific scenario.

Whatever decisions are made with respect to a specific course, the uses of and tradeoffs between various instructional techniques must be re-evaluated frequently. There are many forces for change [11]. These forces evoke such evaluations in order to provide students with educational experiences that will serve them well in the rapidly changing scientific and engineering workplace.

Methods to increase success of underrepresented groups

An article, "Research finds advantages in classes of 13 to 17 pupils", appeared in the April 30, 1999 edition of the *New York Times*. The research discussed in this article involved students in grades K–12, but arguably the results of the study could also apply to college students. The researchers observed that students in these smaller classes had higher grades, better graduation rates, and were more likely to attend college than those from larger classes. Quoting from the article, "The small-class study also found that minority and poor students were helped even more by small classes in some areas than other groups…"

Role models, diversity in the administration, faculty, and student body, and an appropriate balance of competition versus cooperation are just a few of the important factors to consider with respect to encouraging underrepresented groups in math, engineering, and physics.

The effect of learning media on reading, writing, and problem solving

There are tradeoffs between various learning media. For example, Greg Reynolds of New Mexico State University has had good results teaching via a computer projection system in the classroom and the Internet [12]. Students are given hard copies of the class notes, which they can annotate during class as these notes are presented on the computer projection screen. There are links from the notes, which are available on the Internet, to homework problems and solution modules. The instructor also includes computer simulations in the class presentations. Reynolds claims that this more repetitive and visual approach promotes more discussion and interactions between students, and results in improved interest and learning of the material.

Phil Smith, also of New Mexico State University, has presented courses using television and the Internet [13]. At this stage it is unclear what the pedagogical tradeoffs of such a "distance" learning mode are.

Some studies have shown a difference in abilities of students who have taken traditional calculus courses versus students who have taken reform calculus courses. One such study [14] showed a distinct difference in approaches to solving engineering mechanics problems that involve calculus. "Calculus and *Mathematica*® students, who learned calculus with a conceptual emphasis, were found to be more likely to solve problems from a conceptual viewpoint than were the traditional students, who were more likely to focus on procedures." For results of several other studies, see [3, pp. 5–9].

Building theoretical understanding

There are many factors involved in promoting a deep understanding of material covered in freshman and sophomore mathematics courses. In particular, an understanding of the underlying theory is critical. Some of these factors are that [15]

- the students have the prerequisite skills and knowledge,
- the instructor provides clarity of goals and standards,
- there is a perceived relevance of this theoretical course material,
- there is sufficient student practice and instructor feedback,
- testing includes these theoretical topics,
- the instructor has knowledge, preparation, enthusiasm, and empathy,
- there is a reasonable student workload, and
- there exists a match between an instructor's teaching style(s) and a student's learning style(s).

Students should feel some degree of ownership in the course. That is, students should be able to feel that they have developed some ideas or carried out some solutions on their own. This instills confidence. Another advantage is that individuals remember ideas better when they have discovered them on their own [16, Introduction].

Aligning the achieved curriculum and the intended curriculum

Is there a core curriculum? The following is a quote from the spring 1999 Newsletter of the SEFI (European Society for Engineering Education) Mathematics Working Group [17]. "In the 1990s it has come to be recognized that the specialisations of engineering have continued to diversify, possibly to such an extent that a totally common core of mathematics at the degree level is becoming increasingly difficult to define, though there remain large interlocking clusters of common substance."

Traditionally, freshman and sophomore mathematics courses have a diverse student clientele: future mathematicians, engineers, economists and physicists for example. Some students will require a more theoretical approach emphasizing logical structure and symbolic work. Others will require a course that is perhaps more modeling-oriented with emphasis on analyzing the real world and the use of technology. Each course must be made useful to this diverse selection of students. The course must be flexible enough to accommodate diverse student needs. Theory, practical understanding, skill building, and applications all must be considered.

Guiding principles arising from educational research

Regardless of what guiding principles arise from educational research, since many of the researchers who study engineering (mathematics, physics) education are not themselves engineers (mathematicians, physicists), the research is usually published in specialized journals that engineering (math, physics) instructors do not read. However, since much of this work is of direct interest to engineering (math, physics) faculty, there is a need to make this work more accessible to those who can apply it [17].

Should calculus be a laboratory (discovery) course?

If calculus is taught as a laboratory course, instructors can emphasize open-ended problems for which there is more than one solution approach and more than one correct solution. Common sense ideas can be brought into the picture. Students can work in small groups. Working with computer technology, students can solve related problems, each time varying a specific parameter and observing the solution dependence upon that parameter. This use of open-ended problems, group dynamics, and technology enhances teaching and learning for understanding.

Summary

The issues facing undergraduate mathematics are many, and they are complex. There are many schools of thought on the topics that have been raised about instructional techniques [18]. The biggest difficulty that many instructors have with any of these approaches is finding the time required for properly planning and implementing them.

Bibliography

1. D. Hughes-Hallett, A. Gleason, W. McCallum, et al 1998. *Calculus – Single and Multivariable*, Second Edition, New York: Wiley, iii–xi.
2. `http://forum.swarthmore.edu/mathed/calculus.reform.html`, (Aug. 5, 1999).

3. L.R. Mustoe and A.C. Croft, "Motivating engineering students by using modern case studies," *International Journal of Engineering Education (IJEE), Vol. 15*, No. 6, (2000) 469–476.
4. R. Marchand and T.J. McDevitt, "Learning differential equations by exploring earthquake induced structural vibrations: a case study," *IJEE, Vol. 15*, No. 6, (2000) 477–485.
5. D.B. Meade and A.A. Struthers, "Differential equations in the new millennium: the parachute problem," *IJEE, Vol. 15*, No. 6, (2000) 417–424.
6. `http://www.mste.uiuc.edu/murphy/papers/CAlcReformPaper.html`, (Aug. 5, 1999).
7. G.D. Allen, J. Herod, M. Holmes, V. Ervin, R.J. Lopez, J. Marlin, D.B. Meade, and D. Sanchez, "Strategies and guidelines for using a computer algebra system in the classroom," *IJEE, Vol. 15*, No. 6, (2000) 411–416.
8. T.J. Murphy, R.E. Goodman, and J.J. White, "Using the web in multivariable calculus to enhance visualization," *IJEE, Vol. 15*, No. 6, (2000) 425–432.
9. J. Rehig, "Developing web-based courses using an online development guide and templates," *ASEE Computers in Education Journal, Vol. IX*, No. 2, (April-June 1999) 51–55.
10. S. Goldberg, "Bridging the gap between instructor and textbook," *ASEE Computers in Education Journal, Vol. IX*, No. 2, (April-June 1999) 45–50.
11. D. Small, "Core mathematics for engineers, mathematicians, and scientists," *IJEE, Vol. 15*, No. 6, (2000) 432–436.
12. G. Reynolds, and A.T. Hanson, "A computer-based approach for delivery of equation-rich material," Presentation at the 1999 ASEE Conference and Exposition, Charlotte, NC, (June 23, 1999).
13. P. Smith, "Teaching a graduate engineering analysis course by combining TV and the internet," Presentation at the 1999 ASEE Conference and Exposition, Charlotte, NC, (June 23, 1999).
14. C. Roddick, "How students use their knowledge of calculus in an engineering mechanics course," Presentation at the Seventeenth Annual Meeting for the Psychology of Mathematics Education, (October 21–24, 1995).
15. R.M. Felder and R. Brent, "Effective Teaching: A Workshop," Thirty-first Midwest Section ASEE Conference, Tulsa, OK, (April 10–11, 1996) F5.
16. W. Roberts, Editor, 1996. *Calculus: the Dynamics of Change,* Mathematical Association of America, MAA Notes No. 39.
17. N. Steel, Editor, *Newsletter of SEFI (European Society for Engineering Education) Working Group on Mathematics in Engineering Education*, (Spring 1999) 10–11; 15–16.
18. L.R. Mustoe and S. Hibberd, 1995. *Mathematical Education of Engineers*, Oxford: Clarendon Press.

Building Creativity Through Mathematics, Interdisciplinary Projects, and Teaching with Technology

Chris Arney
The College of Saint Rose

Abstract. With the accelerating rate of change in our world, the environment we work in is more unpredictable than ever. College graduates must possess the essential ingredients of creativity to be successful and excel in the twenty-first century. This paper examines the roles of a modeling and inquiry based interdisciplinary core mathematics curriculum in providing opportunities for the development of creativity.

Introduction

One of the primary academic program goals at the United States Military Academy (USMA) is to have graduates think and act creatively. This is a worthwhile goal for any educational program. There are many reasons to have such a goal. The accelerating rate of change in the world is producing an environment that is more unpredictable than ever before with a wider range of options. The creative talents of college graduates are crucial to their success. Our modern world requires people with increased flexibility and adaptability. Successful professional service necessitates college graduates to possess the essential ingredients of creativity: creative thinking, critical thinking, innovative problem solving, intellectual versatility, curiosity, and the ability to deal with ambiguity. The West Point mathematics program, along with its interdisciplinary partners, strives to make contributions to this goal by maintaining a creative environment, implementing an innovative curriculum, and using pedagogy to encourage the development of creativity. This paper outlines the contributions of a core mathematics program and its interdisciplinary activities in these endeavors.

Definitions and Terminology

Imagination is more important than knowledge. — Einstein

Creativity is fundamental to the development of productive graduates. At West Point, the creativity goal supports the overall academic goal "to enable graduates to anticipate and respond effectively to the uncertainties of a changing technological, social, political, and economic world." [3] Creativity is the premier attribute needed to empower college graduates to respond effectively to the uncertainties in our changing world. [6] In the article in this volume by Froyd [4], creativity (and life-long learning) take center stage as course and program goals. Froyd's proposition is that "In the long run, your only sustainable source of competitive advantage is your ability to learn and create faster than others."

Our conception of creativity includes an interrelated set of intellectual skills, personal characteristics, and values. [3] The skills include: original thinking, critical thinking, and innovative problem solving. We define creative thinking as the consideration of a broad range of new, sometimes abstract, ideas and the establishment of new connections and relationships among these ideas. Critical thinking is the performance of careful and exact analysis, ultimately leading to a deeper understanding of an issue. Innovative problem solving is defined as combining knowledge with imagination to produce solutions to problems. The personal characteristics linked with creativity include: versatility, tolerance for ambiguity, willingness to take risks, open-mindedness, confidence, and curiosity. The values connected with creativity include: discipline, perseverance, and responsibility.

Graduates who achieve the creativity goal can confidently confront ambiguous situations. They apply their thinking skills and innovation to solve challenging problems. They are active, independent, and self-directed thinkers and learners. They are able to transfer what they know in one context or discipline to another. They respond successfully to new challenges and situations that require inventiveness. When faced with complex problems, they are able to go beyond traditional approaches to devise more useful and more favorable solutions. They are able to work in collaborative teams, as well as individually, to confront ill-defined situations, to generate new ideas, and to function successfully in different settings.

The USMA core mathematics program was designed with this creativity goal in mind. The program attempts to develop ingredients of creativity in the students. While the essence of mathematical theory is accuracy and precision, mathematics can be applied creatively to solve problems (especially those requiring an interdisciplinary perspective) and used to build the thinking skills, characteristics, and values associated with creativity. This paper provides a brief description of the core mathematics program at USMA and its contributions to the development of creativity.

Core Mathematics Program

An expert problem solver must be endowed with two incompatible qualities, a restless imagination and patient pertinacity.
— Howard H. Eves

Core mathematics includes both the acquisition of a body of knowledge and the development of thought processes. [1, 2] Creativity plays a major role in the thinking component of this program. This intellectual foundation affords opportunities for students to progress as life-long learners, who are able to formulate intelligent questions, research answers, and reach logical conclusions. During their required core mathematics program at USMA, students learn to blend theory and practice. To enhance creative and critical thinking, students study concepts from several perspectives and in many contexts. For example, the program strives for understanding of mathematics from different perspectives: discrete and continuous, linear and nonlinear, deterministic and stochastic, disciplinary and interdisciplinary.

The development of critical thinking skills facilitates the presentation of higher-level concepts. Within the core mathematics classes, concepts are interconnected and applied to problems from various disciplines. The requisite problems, especially the interdisciplinary, open-ended projects assigned in each course, develop student experience in employing new technologies, applying mathematical modeling, and writing in creative ways. In addition, students spend time engaged in experimentation, discovery, and reflection. Through this experience, students develop a curious and experimental disposition and a creative mindset.

Because the core mathematics program at USMA is taken by all students to prepare them for required science and engineering courses, there are seven essential subjects, which are integrated into the four courses. [1] These seven subjects are: differential calculus, integral calculus, vector and multivariable calculus, differential equations, matrix algebra, discrete mathematics, and probability and statistics. West

Point implemented this "7-into-4" curriculum in 1990. Descriptions of the content and the creative component of the four core mathematics courses are as follows:

- First semester (Discrete Mathematical Modeling and an Introduction to Calculus) — discrete mathematics in the form of difference equations, matrix algebra through eigenvalues and eigenvectors, and differential calculus. This innovative course establishes the need for and value of creativity in mathematics. Students are required to analyze data, look for patterns, conjecture solutions, and verify their conjectures. The art of mathematical modeling is introduced as students encounter open-ended, realistic problems from several disciplines.
- Second semester (Calculus I) — integral calculus, differential equations through higher order and systems, and a consolidation of matrix algebra. Once again, conjecturing solutions (undetermined coefficients in differential equations) and using modeling to solve complex, sometimes open-ended problems are major parts of this course.
- Third Semester (Calculus II) — multivariable and vector calculus and discrete mathematics of sequences and series. This course requires students to innovate in order to provide examples of functions meeting certain properties or counter-examples of functions with certain properties. These "inverse problems" require creative thinking and are often initially difficult for students. More open-ended problems are encountered and experience is gained in developing innovative mathematical modeling skills and an interdisciplinary perspective.
- Fourth semester (Probability and Statistics) — descriptive statistics, classical probability, random variables, and hypothesis testing. This course provides students with a new way of thinking and problem solving. Stochastic modeling is introduced and developed. Students confront situations where creativity is needed to collect and display data, to analyze realistic scenarios, and to present their results. A large-scale capstone interdisciplinary project is used to consolidate and refine the mathematical, scientific, and technological knowledge and skill of the students.

Examples

We present examples of activities (homework, exam questions, project components) that we use to build the ingredients of creativity in our students. The first two are questions that show how creativity can be essential in solving open-ended problems and performing mathematical reasoning.

Example 1 (Homework in 1st semester). Given the following difference equation,

$$a(n+1) = 5a(n) - \frac{n\,2^n}{n+3},$$

iterate a few values to reveal a pattern, conjecture a solution, and test your conjecture. Explain why your conjecture was correct or not.

Example 2 (Exam in 2nd semester). Give an example or explain why this can not happen: a function that is differentiable at $x = 3$, but not continuous at $x = 3$.

The next example involves solving an "inverse" problem. Once students learn to find derivatives of functions of two variables, can they reverse the process? There are infinite correct solutions, but this is often difficult for students.

Example 3 (Exam in 3rd semester). Find two different functions of two variables with a derivative vector of [4,-2] at the point (3,7).

The next example looks for innovation in the analysis of a conceptual result. Once again, there are many "correct" answers.

Example 4 (Homework in 4th semester). Given a distinctly bimodal set of data, what are the advantages and disadvantages of using the mean as a measure of location of the data? Can you suggest any (traditional or nontraditional) alternatives?

The following two examples explain two application problems where the art of mathematical modeling was needed to solve and analyze the problems. This is often difficult for first-year students, but they did remarkably well on these two problems.

Example 5 (Project in 1st semester). Using an interdisciplinary perspective, students were required to model the motion of a bridge subject to external forces using a difference equation. Then they were asked to conjecture or test the effects of applying different external forces. Students were required to explain what they discovered and the implications of their results on designing bridges and similar structures. We required and received creative solutions that showed the connections between different external forces and design criteria. Many students discovered the phenomenon of resonance while investigating this problem.

Example 6 (Project in 2nd semester). Students were asked to determine how much grass seed was needed to re-seed an area of campus where no scaled maps were available. As a follow up to this investigation, budget constraints were added and students were asked to determine ways to re-seed the area with a limited amount of seed. Both questions required and produced ingenious solutions.

The final two examples are questions from application projects where creativity helps in the solution.

Example 7 (Project in 3rd semester). Given a contour map of a mountain (i.e., Pikes' Peak), design a path of a hike to traverse from a base camp to the summit. What criteria did you use? Justify that your path meets the criteria? Based on the criteria is your path optimal?

Example 8 (Project in 4th semester). A vital component of a machine has a failure rate of 1 per 20 hours of operation. How many spare components are needed to obtain a 99% probability that the machine can operate for 96 hours? What assumptions did you make to solve this problem?

Student Growth

Because student growth is important in a four-course program, we have developed specific educational threads for all four courses. These threads weave across the objectives in each course to insure student development. In many ways, creativity plays an important role in the development of these threads. Brief descriptions of the five educational growth threads and their major component in the development of creativity are as follows [1]:

- Mathematical reasoning and inquiry: multiple representations (analytic, graphic, numeric), conjecturing, inferences in situations of uncertainty, generalizing concepts from specific examples, providing examples and counterexamples. Creative and critical thinking experiences are provided continually with increasing complexity.
- Mathematical modeling: the art of problem solving, making and identifying assumptions, testing conclusions and sensitivity of assumptions, and solving interdisciplinary problems through collaborative teamwork. These experiences help develop innovative problem solving skills.
- Scientific computing: manipulating and analyzing data, experimenting with parameters, discovering relationships and structures, recognizing capabilities and limitations of computing, understanding information technologies, and performing simple programming. This growth component develops students' imaginations and gives them the opportunity to create their own problem solving tools. New programming languages and nonlinear communication modes, such as html, web-based material, and linked and multi-media textual materials, are perfect tools for learning connections, concepts, and structure in subjects like mathematics.

- Communicating: expressing ideas clearly and effectively, synthesizing concepts, displaying results graphically, and using and understanding mathematical notation and visualization. These activities develop the students' intellectual versatility.
- History of mathematics: appreciating the human endeavor of the development of mathematics, understanding the service role of mathematics, and motivating the further study of the growing and lively discipline of mathematics. These ideas give students a perspective for studying mathematics and the importance of developing their creativity.

Through recent surveys of sample groups, students report confidence and growth in their creative abilities. The statement "I am confident in my ability to demonstrate creative and critical thinking skill in my scientific reasoning" receives 69% agreement from students after their first semester of mathematics [2]. This grows to 72% or greater after completion of the four courses in the core. The disagreement percentage holds at 6% in these surveys, and the remainder of the responses are neutral to this question. These data seem to indicate we are achieving some success in elements of our goal. However, since this question does not capture all the elements of creativity that we desire to measure, new, more detailed, questions will be added to future surveys.

Learning Model

The mind uses its faculty for creativity only when experience forces it to do so.

— Henri Poincaré

Our creativity learning model involves environmental, curricular, and pedagogical dimensions. [3, 5] The foundation for the development of creativity is established through a supportive learning environment. The mathematics program tries to foster this kind of environment. As discussed, our curricular program provides numerous opportunities for creative development. In addition, our pedagogy of interactive and engaging teaching develops the students' thinking processes.

Role of the Environment. West Point educates students in a mathematics program that supports and values curiosity, imagining, exploring, questioning, and risk taking. The program takes advantage of opportunities to present, discuss, and debate ideas in class and to use innovative methods. The advent and use of the internet is a large leap in building and using a creative environment. In any case, factors like section sizes, student population and time, and technologies available have an impact on the environmental climate for developing creativity. As in any institution, the overall academic environment at USMA has elements that foster and at times hinder the development of creativity.

Role of Curriculum. We summarize some of the curricular elements of the USMA mathematics program that make contributions to this goal:

- The four-course program sparks the students' imaginations and develops their curiosity by exposing them to a wide variety of challenging subjects (seven major topics integrated in the four courses).
- The mathematics courses incorporate reasoning, writing, and problem solving that require students to use imagination and innovation.
- Many of the academic projects and computer laboratories are exploratory and require students to cope with ambiguity, to investigate alternatives, and to discover new results.
- The art of technical problem-solving through mathematical modeling is taught throughout the program. Interdisciplinary projects (approximately three projects in each of the four courses) demand the application of intellectual versatility to transfer students' learning from one context or discipline to another. Collaborative group work is the mode for solving these projects. Some of these projects are open-ended and require significant innovation to solve. USMA has been the lead institution in an NSF-sponsored consortium called Project INTERMATH, where faculty write and students solve interdisciplinary live-

Discrete Dynamical Systems and Intro to Calculus	Calculus I — Single Variable Calculus and Differential Equations	Calculus II — Multivariable Calculus	Probability & Statistics
1D Heat Transfer	Flying Strategies	Missile Trajectory	Great Lakes Pollution
Pollution along a River	Terrain Analysis	Laser Guided Munitions	Vehicle Accident Analysis
Chemical Chain Reaction	Aerobic Capacity	Vehicle Collision	Remotely Piloted Vehicle
Great Lakes Pollution	Vibration of an Airplane Wing	The Health Management Organization (HMO)	Model of Dow Jones Industrial Average
SMOG in LA Basin	Air Traffic Control	The Oil Refinery	Hudson River Pollution Data
Car Financing	Clinic Profit Management	Chemistry ABC's	
Making Water in Space	Wheel Suspension Design	Rocket Control	
Water Treatment	Bass Population	The Satellite Problem	
Analysis of Military Retirement Pay	Bungee Cord / Parachute Jumping	Trajectories in 3-space & Least Squares Analysis of Motion Lab Data	
Bridge Vibrations	Telemetry Data Interpretation		
Viral Infection	Real Estate Taxation		
	Road Construction		
	Airport Construction		
	Forest Fire Fighting		
	Water Reservoir Management		
	Cut/fill and Bridge Abutment/Span Computations		
	Railway Headwall Design		
	Earthquake Tower Problem		

Table 1. Titles of example ILAPs recently used in the four core courses at USMA

ly application projects (ILAPs). These projects are co-authored with faculty in partner disciplines and presented to students as challenging, relevant problems to solve in their mathematics course. Examples of recent projects for the four core courses at USMA are given in Table 1.

Role of Pedagogy. Creativity is an indispensable element of the active classroom. One way that faculty establish an interactive classroom is to encourage questioning and discussion. Question-based pedagogy recognizes that questions, not answers, are the driving force in thinking. When answers generate further questions, important thinking and problem solving occur. Students who develop curiosity and ask questions during their academic pursuits are thinking and learning. The mathematics faculty at West Point make use of classroom and homework activities (active learning and group interaction), technology in the form of computer/calculator discovery and experimentation laboratories, and student-relevant reading and discussions, to establish this interactive environment and to promote active learning. Small class sizes (18 or fewer students per class) help instructors insure students are participating fully in all classroom activities. The classroom layout of blackboards all around the room provides opportunities for all the students in the class to be actively involved and to show and explain their work to others. In a typical class, mathematics instructors ask many questions, facilitate lively discussions, and require participation by all students. Student involvement is critical to success in classrooms, computer laboratories, homework, and the development of creativity.

Conclusion

Discovery consists of seeing what everybody has seen and thinking what nobody has thought.
— Albert Szent-Gyorgyi

Mathematics courses can provide opportunities for the development of creativity. The environment, curriculum, and pedagogy of the USMA core mathematics program strive to provide these opportunities. While this may not be a traditional role for mathematics education, programs like the one at West Point have enabled undergraduates to develop their creativity. Students have opportunities to develop their creativity and grow in their capabilities.

References

A college education should equip one to entertain three things: a friend, an idea, and oneself.
— Thomas Ehrlich

1. Arney, D. C., et al., "Core Mathematics at the United States Military Academy: Leading into the 21st Century", *PRIMUS*. Vol. 5 (1995), pp. 343–367.
2. Arney, D. C., et al., 1996. *Math-Science-Technology Goal Program Assessment* (Summary Report). West Point, NY.
3. Forsythe, G., et al., 1997. *Educating Army Leaders for the 21st Century*. West Point, NY.
4. Froyd, Jeffrey E., "First Two Years of Mathematics for Scientists and Engineers", *Proceedings of the Interdisciplinary Workshop on Core Mathematics: Considering Change in the First Two Years of Undergraduate Mathematics,* West Point, NY (1999).
5. Lumsdaine, Edward and Lumsdaine, Monika, 1993. *Creative Problems Solving: Thinking Skills for a Changing World*, New York: McGraw-Hill.
6. Sculley, John, "The Relationship Between Business and Higher Education: A Perspective on the Twenty-First Century", *EDUCOM Bulletin*, Spring issue (1988), pp. 20–24.

Appendices

Interdisciplinary Lively Application Projects (ILAPs)

A: **Mercury in the reservoirs: Water's OK, but don't eat the fish**

B: **Aircraft Flight Strategies**

C: **Analyzing the Safety of a Dam**

Mercury in the reservoirs: Water's OK, but don't eat the fish

Donald Outing
United States Military Academy

Scope and Prerequisites. This activity involves analyzing the biological impact of mercury pollution. Mathematical concepts needed are recurrence modeling and limits. Familiarity with environmental or medical sciences is not required.

The Concern

Public officials are worried about the elevated levels of toxic mercury pollution in reservoirs providing drinking water to the city. They have asked for our assistance in analyzing the severity of the problem. Scientists have known about the adverse affects of mercury to the health of humans for more than a century. The term "mad as a hatter" stems from the nineteenth century use of mercuric nitrate in the making of felt hats.

How does the Mercury get here?

Human activities are responsible for most of the mercury emitted into the environment. Mercury, a byproduct of coal, comes from the smokestack emissions of old, coal-fired power plants. Its particles rise on the smokestack plumes and hitch a ride on prevailing winds. After colliding with the mountain range, the particles drop to the earth.[1] Once in the ecosystem, micro-organisms in the soil and reservoir sediment break down the mercury and produce a very toxic chemical form known as *methylmercury*.

Biological Impact

Mercury undergoes a process known as *bioaccumulation*. Bioaccumulation occurs when organisms (including humans) take in contaminants more rapidly than their bodies can eliminate them, thus the amount of mercury in their bodies accumulates over time. If for a period of time an organism does not ingest any more mercury, its body content of mercury will decline. If, however, an organism continues to ingest mercury, its body content can increase to toxic levels. Humans can eliminate mercury in their sys-

[1] Wayne A. Hall, " Mercury in the reservoirs: Water's OK, but don't eat the fish," *The Times Herald-Record* [Middletown, NY], 11 Jul. 1999, p. 6, cols. 1–4.

tem at a rate proportional to the amount remaining. Methylmercury decays about 50 percent every 65 to 75 days if no further mercury is ingested during that time.[2]

"Safe" Dose

Based on case studies and substantial human and animal data, the U.S. Environmental Protection Agency (USEPA) set the safe monthly dose for methylmercury at 3 micrograms per kilogram (μg/kg) of body weight.[3] This monthly dose is intended to protect the average adult person who weighs 70 kg.

Requirement One

City officials collected and tested twenty (20) samples of bass from each of the affected reservoirs and have provided us with the data. All fish tested were contaminated. The mean value of methylmercury in the fish samples was 0.09 parts-per-million (ppm) or micrograms per gram (μg/g). The average weight of the fish was 1.5 kg.

If each person adheres to the fish consumption restrictions as published in a public health advisory and consumes no more than one fish per month, construct a recurrence model for the amount of methylmercury that will bioaccumulate in the average adult person. Use your model to determine the maximum amount of methylmercury the average adult human will bioaccumulate in their lifetime.

Requirement Two

What are the primary assumptions you made to develop your model in Requirement One (state at least two)? Revise one of your assumptions so that your model changes. Write a new model that differs from the original model due to the new assumption. How does this affect your answer to requirement one?

Requirement Three

The toxicologist at the local hospital provides you with the following information regarding the human health effects of mercury toxicity: Toxicity is defined using a term called LD50 — a scientific term which literally translates to "lethal dosage, 50th percentile." Simply put, LD50 is the dosage at which 50% of the humans exposed to a particular chemical will die. In our case, the term applies to oral lethal dosage and is expressed in milligrams per kilogram (mg/kg) of body weight. The LD50 for methylmercury is 50 mg/kg.

According to your model from Requirement One, will the reservoir advisories protect the average adult from reaching the LD50? What is the maximum number of fish the average adult can safely eat per month without ever bioaccumulating a lethal dosage of methylmercury?

---

[2] According to the U.S. Geological Survey, the half-life of mercury in the human body is on average 70 days.

[3] The USEPA's reference dose or "safe" daily dose (RfD) is actually 0.1 micrograms per kg of body weight per day.

Aircraft Flight Strategies

David Arterburn, Michael Jaye, Joseph Myers, and Kip Nygren
United States Military Academy

Scope and Prerequisites. This activity involves analyzing flight strategies through relationships containing derivatives. Mathematical concepts needed are modeling with derivatives, numerical integration, analytic integration, and graphical analysis. Familiarity with fluid mechanics is not essential; all required relationships are presented in the Background Material section at the end.

Introduction

Three important considerations in every flight operation are the altitude (possibly variable) at which to travel, the velocity (possibly variable) at which to travel, and the amount of lift that we choose to generate (at the expense of fuel consumption—again possibly variable) during the flight. It turns out that when planning a flight operation, one cannot just choose any desired value for each of these three quantities; they are dependent upon one another. We can relate these three quantities through a set of equations known as the Breguet (pronounced *brĕ-ga'*) Range Equations. These equations are derived in the Background Material section given below. Deriving these equations shows that once we decide to choose constant values for any two of altitude, lift coefficient, and velocity, the third is automatically determined. Thus there are three basic independent flight strategies: constant altitude/constant lift coefficient, constant velocity/constant altitude, and constant velocity/constant lift coefficient. Requirement 1 asks you to analyze how the third quantity must vary under each of these flight strategies.

Commercial flight operations are generally conducted at constant velocity and constant lift profile in order to save fuel. However, in special operations (e.g., search and rescue, military operations) there are often other considerations that override cost efficiency, and thus dictate the choice of a different flight strategy. When several aircraft are in the air at the same time, especially both outbound and inbound, safe airspace management often dictates flights at constant specified altitudes. Requirement 2 asks you to more closely analyze which flight strategy may be most appropriate for different missions. Thus unlike many commercial operations, some airspace planners must be prepared to operate under any of several different flight strategies.

The following scenarios demonstrate how different techniques of single variable calculus can assist in analyzing the governing equations to yield important information about flight operations.

Scenario

You are the pilot on a search aircraft, and among the many things for which you are responsible, you have to determine within what radius your plane can safely service a search mission, how long it can stay in the area, and when it must return for refueling.

Now, an interesting aspect of your job is that, at times, some of the instruments malfunction. This forces you to double-check your instruments' accuracy through other means, or to rely on these other means to plan your plane's flight. In this project you are going to answer several questions about the flight of your craft based primarily on your plane's fuel consumption. (Your fuel gauge is known to be working).

Strategy 1: Flying at Constant Velocity and Constant Lift Profile

Range Equation. You can answer questions regarding how far the plane can travel by relating the distance traveled by the plane to the weight of fuel that it consumes. Assume that you fly at constant velocity and with a constant coefficient of lift (thus, you increase altitude over time as your plane gets progressively lighter). From our knowledge of fluid dynamics, we have the following relationship (this and all following relationships are derived in the Background Material section at the end):

$$\frac{dx}{dW} = -\frac{V}{c}\frac{C_L}{C_D}\frac{1}{W},$$

where x = distance traveled, W = weight, V = velocity, c is the coefficient of fuel consumption ($c = 0.3700$ lbs. of fuel/hr/lb thrust), and the ratio C_L/C_D is 3.839 for constant lift coefficient. Thus, the distance traveled, x, is given by:

$$x = -\frac{V}{c}\frac{C_L}{C_D}\int_{W_{start}}^{W_{finish}} \frac{1}{W}dW .$$

Example 1. You take off weighing 40,434 lbs (this weight includes fuel) and you travel at 347.5 mi/hr. You arrive at the search area weighing 36,434 lbs. By use of a numerical integration technique, with an increment size of 1000 lbs in your partition, estimate the distance you have traveled. Does your answer depend on your increment size?

Solution. This requires us to numerically evaluate the integral $-3605.8\int_{40,434}^{36,434}\frac{dW}{W}$, which we rewrite as $3605.8\int_{36,434}^{40,434}\frac{dW}{W}$. We use the trapezoidal rule

$$x = 3605.8*(.5*f_0*\Delta W + \sum_{i=1}^{n-1} f_i*\Delta W + .5*f_n*\Delta W)$$

to evaluate the integral, with $f(W) = 1/W$ and $\Delta W = 1000$. Substituting for f yields:

$$x = 3605.8*(.5*36,434*\Delta W + \sum_{i=1}^{n-1} {}^{1}\!/_{W_i}*\Delta W + .5*40,434*\Delta W),$$

where W_i is the value of W in the i'th subinterval (equal to $W_i = 36,434 + (i-1)*(40,434 - 36,434)$). This technique is implemented in the following spreadsheet:

Initial W:	36434		
Final W:	40434		
Intervals:	4		
Delta W:	1000	Distance:	375.6537
W	f(W)	Partial Sum	
36434	2.74469E-05	0.013723	
37434	2.67137E-05	0.040437	
38434	2.60186E-05	0.066456	
39434	2.53588E-05	0.091815	
40434	2.47317E-05	0.10418	

This yields a distance traveled of 375.6 miles. We look to the next example to better answer the question "is the calculated range a function of increment size?"

Example 2: Refine your estimate by increasing the number of partitions. What appears to be the limit as the number of partitions increases without bound?

Solution: Repeating the above process for differing numbers of subintervals yields the following sequence of values for the distance traveled:

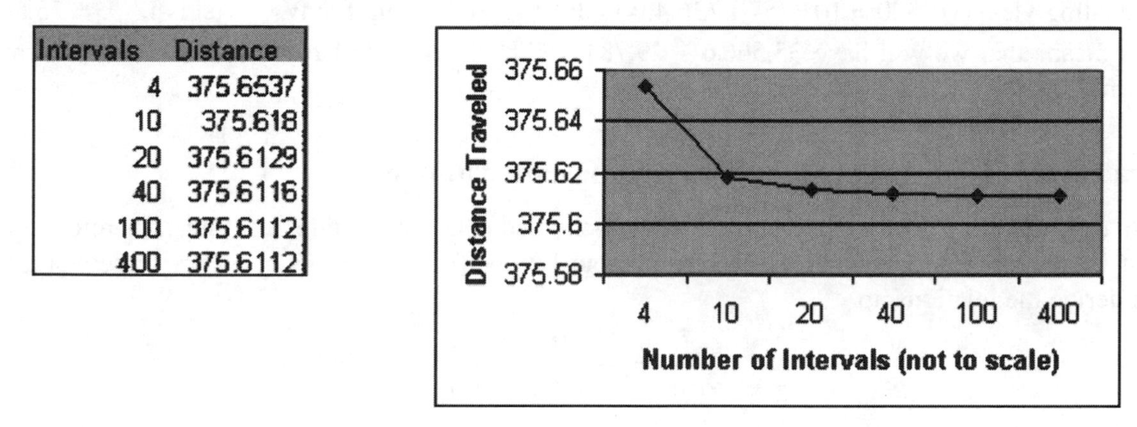

Intervals	Distance
4	375.6537
10	375.618
20	375.6129
40	375.6116
100	375.6112
400	375.6112

The calculated range appears to be a monotonically decreasing function of the number of subintervals (or conversely, a monotonically increasing function of increment size). This also appears to be a convergent sequence with a limit of approximately 375.6 miles. Note how few terms are required (in this case) to converge very close to the apparent limit of the numerical integration scheme.

Example 3: Now evaluate the definite integral to find the distance traveled.

Solution: Evaluating the definite integral, which is easy to do for this simple integrand, yields

$$3605.8 \int_{35,434}^{40,434} \frac{dW}{W} = 3605.8 * \ln(W) \Big|_{36,434}^{40,434} = 375.6112$$

miles. This is in excellent agreement with the numerical solution above.

Endurance Equation: To determine how long you can stay in the search area with a given amount of fuel, we need to relate the time t to the fuel consumption. With the help of some equations from our fluid dynamics background, we find that, if we assume that we are loitering at a constant velocity, V, and a constant lift coefficient C_L, we have

$$\frac{dt}{dW} = \frac{\frac{dx}{dW}}{\frac{dx}{dt}} = -\frac{1}{c}\frac{C_L}{C_D}\frac{1}{W}.$$

Thus, t, the loiter time, is given by:

$$t = \int_{W_{begin}}^{W_{end}} -\frac{1}{c}\frac{C_L}{C_D}\frac{1}{W}\,dW = -\frac{1}{c}\frac{C_L}{C_D}\int_{W_{begin}}^{W_{end}} \frac{1}{W}\,dW.$$

Example 4*:* You arrived at the search area weighing 36,434 lbs. You have to search for 15 minutes (0.25 hour). How much fuel will you have for your return trip assuming that the plane weighs 29,784 lbs without fuel, but with supplies you have for stranded people?

Solution. Substituting into the endurance equation yields

$$0.2500 = -10.3757 \int_{36,434}^{W_{final}} \frac{dW}{W},$$

which we rewrite as

$$0.2500 = 10.3757 \int_{W_{final}}^{36,434} \frac{dW}{W},$$

Evaluating yields $0.2500 = 10.3757(\ln(36{,}434) - \ln(W_{final}))$. Solving for W_{final} yields $W_{final} = 35{,}566.6$ lb. This means that we will have 35,566.6 – 29,784 = 5782.6 lbs of fuel remaining when we are ready to return.

Strategy 2: Flying at Constant Velocity/Constant Altitude

You are required to return home at constant velocity and constant altitude. You must, therefore, decrease your lift as your plane lightens by decreasing your lift coefficient. It turns out, after some work, that we can derive the relationship

$$\frac{dx}{dW} = -\frac{V}{c\bar{q}SC_{D_0}} \frac{1}{1+aW^2},$$

where $a = 2.330 \times 10^{-11}$, $C_{D_0} = 0.03700$, $\bar{q} = 541.894$, $S = 506.0$ ft^2 (the surface area of the wing), and $c = 0.3700$ lbs of fuel/hr/lb thrust. Thus, the distance traveled, in miles, is given by:

$$x = \int_{W_{depart}}^{W_{arrive}} -\frac{V}{c\bar{q}SC_{D_0}} \frac{1}{1+aW^2} dW = -\frac{V}{c\bar{q}SC_{D_0}} \int_{W_{depart}}^{W_{arrive}} \frac{1}{1+aW^2} dW.$$

Example 5. Your mission is complete, you've found the missing people and dropped supplies to them, and you find yourself 478.0 miles away from the airfield. You will return to the field at a constant velocity, $V = 460.4$ mi/hr, and at a constant altitude. Can you make it home on 4500 lbs of fuel? If so, then how much fuel do you have remaining when you do arrive? If not, then how much additional fuel would you need? Your craft weighs 24,959 lbs when empty of fuel and supplies.

Solution. Substituting into the constant velocity/constant altitude equation yields

$$x = -.12265 \int_{29,459}^{24,959} \frac{dW}{1+2.330\times 10^{-11} W^2}.$$

Note that we have the freedom here to choose any integration technique (numerical, analytic, Computer Algebra System (CAS)) that we desire. We find symbolically that

$$x = -.12265\left(2.330\times 10^{-11}\right)^{-1/2} \tan^{-1}\left(\sqrt{2.330\times 10^{-11}}W\right)\Big|_{29,459}^{24,459},$$

or evaluating numerically that $x = 542.546$ miles. Therefore, we will make it home with 542.5 – 478.0 = 64.5 miles to spare.

Strategy 3: Flying at Constant Altitude/Constant Lift Coefficient:

We have discussed two flight strategies, namely flight at constant velocity/constant lift coefficient, and flight at constant velocity/constant altitude. A third strategy is constant altitude/constant lift coefficient. Now, constant lift coefficient will require you to slow down over time as your plane lightens (otherwise your plane will climb). It turns out for this strategy that we can derive the relationship

$$\frac{dx}{dW} = -\frac{1}{c}\sqrt{\frac{2}{\rho S}}\frac{\sqrt{C_L}}{C_D}\frac{1}{W^{1/2}}.$$

So the distance that you can travel, in miles, is given by

$$x = \int_{W_{start}}^{W_{end}} -\frac{1}{c}\sqrt{\frac{2}{\rho S}}\frac{\sqrt{C_L}}{C_D}\frac{1}{W^{1/2}}dW = -\frac{1}{c}\sqrt{\frac{2}{\rho S}}\frac{\sqrt{C_L}}{C_D}\int_{W_{start}}^{W_{end}}\frac{1}{W^{1/2}}dW$$

where $\rho = 0.002377$ slug/ft^3 (air density) and $\sqrt{C_L}/C_D = 9.997$.

Requirements

1. Use the Breguet range equations to determine the following. In each case, explain why your answer is intuitively plausible.
   a. For a constant altitude/constant lift coefficient flight operation, how must the velocity of the aircraft vary during the flight?
   b. For a constant velocity/constant altitude flight operation, how must the lift coefficient of the aircraft vary during the flight?
   c. For a constant velocity/constant lift coefficient flight operation, how must the altitude of the aircraft vary during the flight?
2. If you have only a limited amount of fuel on board, which of the three flight strategies allows you to travel the furthest? Is any one of the three always best? Is any one of the three always the worst?
3. The Voyager was the first aircraft successfully flown non-stop around the world. How do you think the Breguet equations (along with other design considerations) played a role in the design of this unique aircraft for this very specialized mission?

Background Material: Derivation of the Breguet Range and Endurance Equations

1. Mathematical Model

Lift (L) = Weight of the aircraft (W) (by Newton's second law, assuming no or negligible vertical acceleration)

Thrust (T) = Drag on the aircraft (D) (by Newton's second law, assuming no or negligible horizontal acceleration)

Velocity (V) = dx/dt (where x is the position of the plane at time t)

$-\frac{dW}{dt} = cT$ (loss of weight, all due to fuel consumption, is directly proportional to the thrust produced; c is the specific fuel consumption in units of lbs fuel/(hr x lbs thrust))

2. Definitions:

Coefficient of lift: $C_L = \frac{L}{\overline{q}S}$

Coefficient of drag: $C_D = \frac{D}{\overline{q}S}$

$C_D = C_{D_0} + KC_L^2$, where $\overline{q} = \frac{1}{2}\rho V^2$, ρ = air density, S = wing area, and C_{D_0} and K are constants

3. Derived Relationships:

$$\frac{L}{D} = \frac{C_L}{C_D}$$

$$T = D = W\frac{D}{L} = W\frac{C_D}{C_L}$$

$$V = \sqrt{\frac{2W}{\rho S C_L}}$$

4. Range Equation for Constant Altitude (ρ constant) and constant C_L:

$$-\frac{dW}{dt} = -\frac{dW}{\frac{dx}{V}} = cT, \text{ or } -\frac{dW}{dx} = \frac{cT}{V} \text{ and } \frac{dx}{dW} = -\frac{V}{cT}$$

By substituting for V:

$$\frac{dx}{dW} = -\frac{1}{c}\sqrt{\frac{2W}{\rho S C_L}}\frac{C_L}{C_D}\frac{1}{W} = -\sqrt{\frac{2}{\rho S C_L}}\frac{C_L^{1/2}}{C_D}\frac{1}{W^{1/2}}$$

5. Range Equation for Constant Velocity and Constant C_L:

$$\frac{dx}{dW} = -\frac{V}{c}\frac{C_L}{C_D}\frac{1}{W}$$

6. Range Equation for Constant Velocity and Constant Altitude:

$$\frac{dx}{dW} = -\frac{V}{cT} = \frac{V}{cD}$$

Substituting for Drag, where

$$D = \bar{q}SC_D = \bar{q}S\left(C_{D_0} + KC_L^2\right) \text{ and } C_L = \frac{W}{\bar{q}S}$$

yields:

$$\frac{dx}{dW} = -\frac{V}{c\left(\bar{q}SC_{D_0} + \frac{KW^2}{\bar{q}S}\right)} = -\frac{V}{c\bar{q}SC_{D_0}}\frac{1}{1+aW^2}, \text{ where } a = \frac{K}{\bar{q}^2S^2C_{D_0}}$$

7. Endurance Equation for a Jet Aircraft at Constant C_L:

$$\frac{dt}{dW} = -\frac{1}{cT} = -\frac{1}{c}\frac{C_L}{C_D}\frac{1}{W}$$

References

1. Anderson, John D., *Introduction to Flight*, 3d Ed., New York, McGraw-Hill, 1989.
2. Millikan, Clark, B., *Aerodynamics of the Airplane*, New York, Wiley, 1941.
3. Perkins, Courtland D., and Robert E. Zhage, *Airplane Performance, Stability, and Control*, New York, Wiley, 1949.

Analyzing the Safety of a Dam

Steve Horton, Dennis Day, Andre Napoli,
Brett Barraclough, Gerald Hansler, Joe Samek
United States Military Academy

Scope and Prerequisites. This activity involves analyzing statistical data. Mathematical concepts needed are modeling, curve fitting, elementary calculus, and statistics. Familiarity with basic concepts of statics in elementary physics is also helpful.

Situation

You are in Thailand as part of a Civic Action Team providing assistance. Your company recently has initiated the construction of a school and several roads in the area. The US Ambassador to Thailand contacts you and requests that you help her investigate a recently rebuilt dam in the area. The ambassador is meeting with the President of Thailand in two days, and wants the results of your investigation tomorrow. The President has been informed that the new dam may not have been designed correctly given the historical data available regarding water height. The major concern seems to be that the dam will tip over, potentially endangering thousands in the flood plain below. A secondary concern is that the average maximum water height might be too low in some months to support vital irrigation projects. Finally, the results of compressive strength tests performed on sample cylinders of concrete used to make the dam have recently returned from a lab, and the President wants to know if the concrete used was "up to code." Your report should address each of these concerns. The ambassador's aide provides you with drawings (Figures 2 and 3), some concrete sample data (Table 1), and some water height data (Table 2). The ambassador wants a detailed analysis that includes any assumptions you make, an explanation of all the mathematics and science you use, and recommendations. She also wants to know the details regarding any computer software you use.

Additional Information

1. The dam has a footer that prevents it from sliding in any direction. Assume that it can only fail by tipping forward. Also assume that the operators of the dam cannot significantly affect water height. The historical water height data provided (see Table 2) is based on the same type of dam.
2. The density of the concrete is 145 lbs/ft3.
3. If the mean maximum water height for any month is below 20 feet, irrigation projects will be adversely effected.
4. Concrete must have a compressive strength of 3000 psi to be considered "up to code."

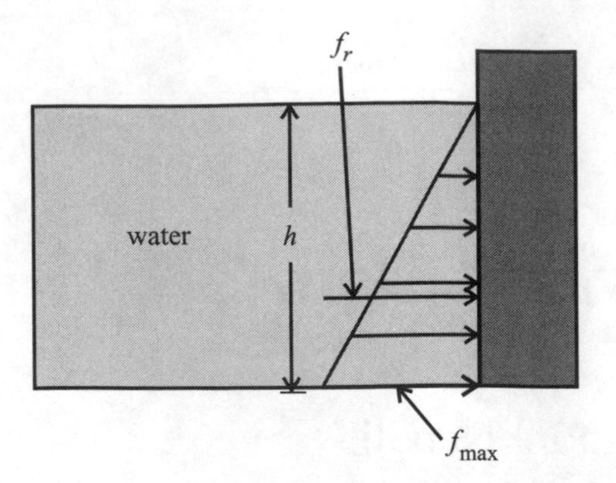

Figure 1. Hydrostatic Force of Water on a Plane Surface

5. The hydrostatic force of water on the vertical face of the dam is described in Figure 1. The maximum linear hydrostatic force over the plate area is given by

$$f_{\text{max}} = \gamma h a$$

where γ is the specific weight of water, h is the height of the water and A is the surface area in contact with the water. The resultant linearly distributed force, f_r, can be found using calculus.

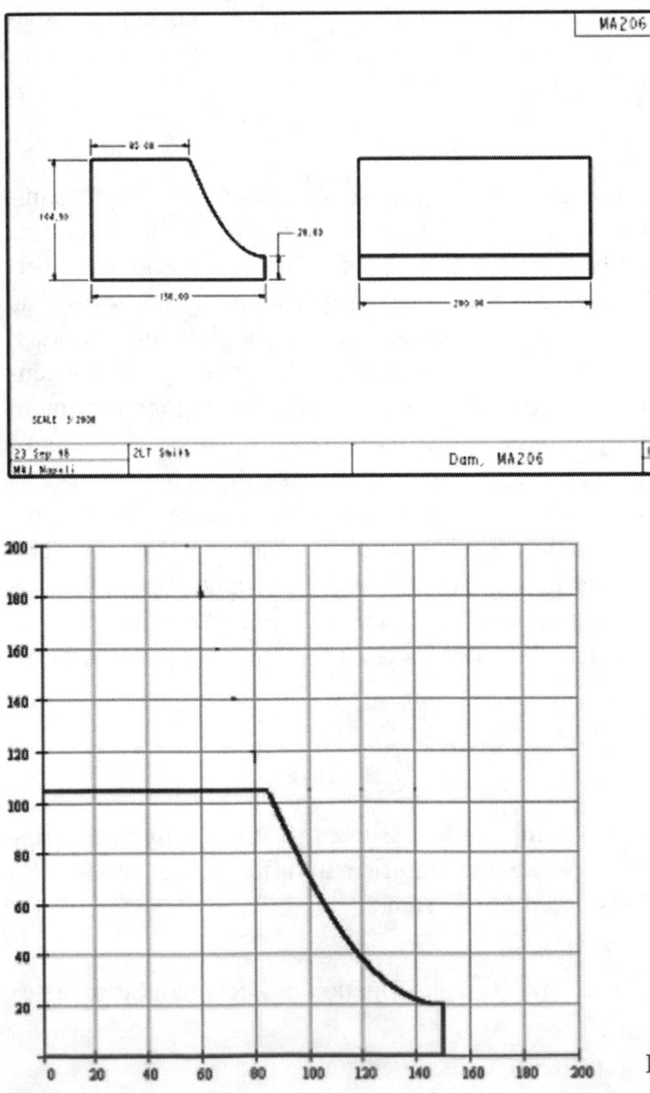

Figure 2. Side and Front View of Dam

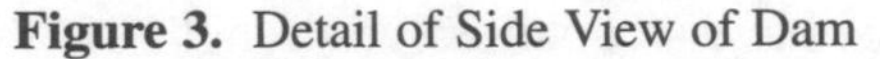
Figure 3. Detail of Side View of Dam

3017	2984	3046	3095	3066	3139	3064	2987	2964	3072

Table 1. Concrete Compressive Strength Data

Concrete Compressive Strength Data

The following are the concrete compressive strength data. Table 1 gives the compressive strength, in pounds per square inch, of 10 test cylinders of concrete used to make the dam.

Water Height Data

The following are the river height data. Each column contains data from a specific month of some year. Each entry in the table contains the monthly maximum water height (in feet) for some previous year. for example, your first column contains the maximum water height in the month of January for 32 previous years. Note that you have more observations for some months than you do for others.

JAN	FEB	MAR	APR	MAY	JUN	JUL	AUG	SEP	OCT	NOV	DEC
69.5	51.4	40.2	36.4	11.3	30.5	42.7	60.5	77.7	97.1	87.2	79.5
70.5	49.5	37	31.1	16.9	31.6	40	60.4	81	103.9	92.6	84.4
71.5	54.6	42.6	24.8	15.6	30.1	41.6	61.8	80.5	99.9	89.8	89.6
71.4	52.5	41	27.7	12.9	26.7	44.7	61.2	81.9	100.2	93.4	82
71.1	49.8	43	31	14.9	31.9	43.4	60.4	80.6	105	90.6	72.8
68.7	51.5	37.4	36	22.3	20.1	47.3	58.3	79	103	88.2	79.9
71.7	46.4	39.9	34.9	15.5	24.5	39.5	60	81.8	95.8	88.9	86.8
72.9	51.8	38.6	31	15	20.5	42.2	60.5	80.5	102.4	87.3	77.9
68.8	45.1	37.7	24.8	19.7	21.1	43.2	62	82.7	106.2	90.4	82.2
71.9	47.8	42.1	31	22.8	24.8	35.7	59.3	82.8	103	89.4	83.7
70.1	42.5	40.1	25.9	19.5	19.5	44.1	60.3	77.5	104	87.4	76.2
68.8	53.2	39	31.4	15.6	19.9	39.8	61.3	78.7	101.9	91	84.8
72.3	53	41.2	25.3	20.3	21.5	41.1	58.9	79.3	104.1	87.5	75.8
71.8	52.5	39.3	29	16.5	28.2	36.4	60.8	78.2	103.4	89.5	88.8
68.9	47.9	40.9	34.6	19.1	18.3	39.5	63	80.4	102	90.4	74.9
70.9		41.7	36.5	21.6	27.1	37	62.5	77	106.3		80.9
71.3		40.6	35.1	15.8	37.5	39.2	61.7	77	106.1		79.9
71.3		37	31.5	6.8	24.8	38.8	59.2	76.9	101.4		85
72.7		42.2	38.1	13.4	26.6	33.4	59.6	79.4	102.5		78.4
69.3		39.9	31.4	15.9	20.5	43.3	61.7	76.9	104.8		87.5
72		42.4	30.8	22.2	32.7	38.8	59.8	76.3	98.8		73
72.5		41.6	30.2	18.2	21.2	40.1	58.4	76.2	98.7		82.7
71.7		40.3	25.4	12.9	23.3	38.3	62.5	77.8	103.9		82.7
70.6		-888	27.3	20.3	24.9		60.8	78.3	92.1		79.5
70.8		39.4	25	17.1	18		59.6	80.9	97.9		76.4
70		38.4	28		28.6		60.3	76.7	103.5		71.9
72.7		42.7	26.9		19.3		58.3	80.4	101.6		81.2
72.9		38.6	30.8		21.1		59.6	78	101.5		80.3
72.4		40.2	27.7		23.9		61.3	78.3	97.2		73.9
70.9		41.5	27.1		26		59.9	83	104.4		83.2
70.3		41.9	38.5		19.3		59.5	82.1	98.6		78.6
72.9		40.6	27.8				61.6	79.3	101		
		43.8	24.2					81.7	108.2		
		39.8	30.1						105.3		
		38.2	24.4						98		

About the Editors

Chris Arney graduated from the United States Military Academy in 1971. He went on to receive his PhD in mathematics from Rensselaer Polytechnic Institute. Most of his military career was spent as a mathematics professor at West Point. He retired from the Army in 2001 as a Brigadier General. He currently serves as the Dean of the School of Mathematics and Sciences at the College of Saint Rose in Albany, New York. He has taught undergraduate mathematics for 20 years, and is the author of over 20 books. His technical areas of interest include applied mathematics, operations research, computer science, and the history of mathematics and science. His teaching interests include the use of technology, the use of interdisciplinary problems, and the revitalization of curricula to improve undergraduate teaching. Chris has been very active in the area of curriculum reform. He is the Director of the Interdisciplinary Contest in Modeling, an annual undergraduate competition involving student teams from schools around the world. In addition to his work in mathematics, he has written two expository works under the pen name of Alan Firstone—a novel *Son of the Silvery Waters* and a nonfiction work entitled *The Annotated Rose.* He has recently completed a work entitled *West Point's Scientific 200: Celebration of the Bicentennial. Biographies of 200 of West Point's Most Successful and Influential Mathematicians, Scientists, Engineers, and Technologists.*

Don Small received his PhD from the University of Connecticut. He taught mathematics at Colby College for 23 years before joining the Department of Mathematics at the United States Military Academy in 1991. He is active in the calculus and college algebra reform movements—developing curricula, authoring texts, and leading faculty development workshops. Don has served two terms as Governor of the Northeast Section of the Mathematical Association of America (MAA), and has been a long-term member of the MAA Calculus Reform in the First Two Years committee. His interests are in developing curricula that focus on student growth while meeting the needs of partner disciplines, society, and the workplace. He is the author of several textbooks: *A Unified Introduction to Linear Algebra, Models, Methods and Theory* (with Alan Tucker), *Calculus: An Integrated Approach,* and *Explorations in Calculus with a Computer Algebra System* (both with John Hossack), *Exploring Calculus with Math T/L* (with Douglas Child) and *Contemporary College Algebra: Data, Functions, Modeling.* With co-editors Carl Leinbach, Joan Hundhausen, Arnold Ostebee, and Lester Senechal, he edited the *Laboratory Approach to Teaching Calculus*, volume 20 in the MAA Notes Series.

4987